DICTIONARY OF SCIENCE

Mahender Yadav

ANMOL PUBLICATIONS PVT. LTD.
NEW DELHI - 110 002 (INDIA)

ANMOL PUBLICATIONS PVT. LTD.
Regd. Office: 4360/4, Ansari Road, Daryaganj,
New Delhi-110002 (India)
Tel.: 23278000, 23261597, 23286875, 23255577
Fax: 91-11-23280289
Email: anmolpub@gmail.com
Visit us at: www.anmolpublications.com

Branch Office: No. 1015, Ist Main Road, BSK IIIrd Stage
IIIrd Phase, IIIrd Block, Bengaluru-560 085 (India)
Tel.: 080-41723429 • Fax: 080-26723604
Email: anmolpublicationsbangalore@gmail.com

Dictionary of Science

First Edition, 1998

Reprint, 1989, 1990, 1991, 1992

Second Revised Edition 1993

Reprint, 1994, 1995, 1996, 1997, 1998, 2000, 2002, 2003, 2004, 2006, 2008, 2010, 2012

PRINTED IN INDIA

Printed at Suman Printers, New Delhi

Preface

It is a matter of great satisfaction that the first edition of this dictionary got exhausted within a short period. Now I am presenting the new revised and enlarged second edition. This has been revised to make it up-to-date.

This dictionary of science is an attempt to put the terms and definitions used in various branches of science in a compact form for ready reference. Most of the terms commonly used have been listed and an attempt has been made to make the terms as much clear as is possible. Very simple language has been used in most of the terms except where it has not been possible due to the scientific definitions of the terms. S I units have been used but at some places other measurement units have also been included. Some terms included in this dictionary might have become obsolete but these have been included for their historical importance.

An attempt has been made to make a balance between the space available while explaining a term. However, every effort has been made to explain various terms fully.

A number of diagrams and pictorial representations have been used to make the terms very clear or to help in understanding a term.

In compiling a dictionary of this kind, it becomes necessary to draw upon the work of many authorities and seek the advice of colleagues to all of whom the editor is deeply indebted.

Editor

A

A. The symbol for relative atomic mass. The A value for an element is a measure of the mass of an average atom of the element compared to the mass of an atom of the isotope carbon-12.

Ab. Used as a prefix to the names of practical electric units (*e.g.* *(ampere, volt)* to indicate the corresponding unit in the *electromagnetic system* (*e.g.*, abampere, abvol See *absolute units*).

Abacus. A frame containing parallel wires carrying beads used for computation.

Abbe Condenser. It is an optical *condenser* and is used in *microscopes.* It consists of two or three *lenses* having a wide *aperture*. It has been named after Ernst Abbe (1840-1905).

Abdomen. In mammals, the part of the body separated from the thorax by the diaphragm containing stomach, liver, intestines, etc.

Aberration (astr.) Refers to variation in the apparent position of a *star* or other heavenly body. It is due to the motion of the observer with the *Earth*.

Abberration Chromatic. Refers to the formation, by a *lens*, of an image with coloured fringes. As the *refractive index of glass* is different for *light* of different *colours* the light is dispersed into a coloured band. The effect can be corrected by the use of *achromatic lenses*.

Aberration, spherical. Refers to the distortion of the image formed by a *lens* or *mirror*. It arises due to different rays from any one point of the object making different angles with the line joining the point to the *optical centre* of the lens or mirror and coming to a *focus* at slightly different positions.

Abiogenesis. It is a hypothetical process by which living organisms are created from non-living matter ; autogenesis.

Abiotic. Absence of living organisms.

Abiotic Environment. Factors that influence living organisms, both physical and chemical.

Abort. To halt a processing activity in a computer before some planned conclusion has been reached.

Abscission. Refers to liberation of fungal spore by breakdown of part of its stalk or shedding of bark, leaf, flower or fruit by a plant. It is usually due to decline in auxin level in that plant.

Abrasive. It is a substance that is used for rubbing or grinding down surfaces ; *e.g., emery.*

Abscissa of a point P. It is thé portion of the *x* axis that lies between the *origin* and a point where the line through *P* parallel to the *y* axis cuts the *x* axis.

Absolute. Not relative ; independent *absolute zero* of temperature is different from zero on an arbitrary scale such as the *Celsius temperature* scale.

Absolute Alcohol. *Ethanol* that contains not less than 99% pure ethanol by weight.

Absolute Expansivity. The true *expansivity* of a liquid not relative to the containing vessel. It is equal to the sum of the relative or apparent expansivity of the liquid and the volume expansivity of the containing vessel.

Absolute Humidity. The amount of *water vapour* present in the *atmosphere.* It is defined in terms of the number of *kilograms* (or grams) of water that is present in one cubic *metre* of air.

Absolute Temperature (Kelvin scale K). Temperature measured in degrees celsius (centigrade) but from absolute zero (and not from) freezing point of water.

Absolute zero. The lowest temperature theoretically possible ; OK = —273.15°C = —459.67°F.

Absorptance α. It is the ratio of the radiant or luminous *flux* absorbed by a body to the fiux falling on it. The absorptance in case of a *black body* is 1.

Absorptiometer. It is the apparatus used for determining solubility of gases in liquids.

Absorption. It is the process by which gas is taken up by liquid or solid or liquid is taken up by solid.

Absorption Coefficient α. 1. It is the ratio of the sound energy absorbed at a boundary to the sound energy falling on it.

Absorption Edge. It is the *wavelength* at which a discontinuity appears in the intensity of an X-ray *absorption spectrum.*

Absorption of Gases. The *solution* of gases in *liquids* or absorption of gases by *solids* when the gas permeates the whole bulk of the solid.

Absorption of Radiation. *Radiant energy* is partly reflected, partly transmitted, and partly absorbed by the surface upon which it falls. Dull black surfaces are the best absorbers and also the best radiators.

Absorption Spectrum. A *spectrum* (which consists) of dark lines or bands obtained when the *light* from a source, is passed through a gas into a *spectroscope*. When the absorbing medium is in the *solid* or *liquid state* the spectrum of the transmitted light shows broad dark regions.

Absorptivity. Of a surface it is the fraction of the *radiant energy* incident on the surface that is absorbed.

Abstriction. Release of spore from its stalk by constriction of the stalk.

Abundance. 1. It is the ratio of the number of *atoms* of a particular *isotope* in a mixture of isotopes of an *element*, to the total number of atoms present. Sometimes, it is expressed as a percentage, *e.g.*, the abundance of U-235 in natural uranium is 0.71%. 2. It is the ratio of the mass of an element present in the earth's crust to the total mass of the earth's crust. For example, the abundance of silicon in the earth's crust is 28%.

Abyssal. It denotes a zone of the ocean, or an organism that inhabits it. It is at a depth of about 2000 metres, where very little light penetrates.

Acceleration *a*. It is the rate of change of *velocity* (*v*) or *speed*. It is measured in ms^{-2}.

Acceleration of Free Fall *g*. Acceleration due to gravity. The *acceleration* of a body falling freely in a *vacuum*; standard accepted value is 9.806 65 ms^{-2} (32.174 ft s^{-2}).

Accelerator (chem.). A substance which increases the rate of a *chemical reaction* (*i.e.*, a *catalyst*) *e.g.*, in the manufacture of *vulcanized rubber.*

Accelerator (phys.) A machine used for increasing the *kinetic energy* of charged particles (*e.g., protons electrons, nuclei*) by accelerating them in *electric fields.*

Accelerometer. It is an instrument used to measure *acceleration,* especially the acceleration of an aircraft or *rocket.*

Acceptor. An imperfection in a *semiconductor*. It causes *hold* conduction.

Accomodation. Refers to the adjustment of eyes to distinct vision at different distances.

Accretion. It is the continued growth or increase in size by external addition.

Accuracy. This indicates how close a measurement is to the correct value.

Acetal. It is an *organic compound* having a general formula $RCH(OR')_2$. $CH_3CH(OC_2H_5)_2$, 1, I-diethoxyethane, known as acetal boils at 104°C and used as a *solvent*, in perfumes, and in organic synthesis.

Acetate Plastics. *Plastics* made from *Cellulose* acetate.

Acetic acid (VH_3COOH). Colourless, corrosive liquid having pungent smell boiling point 118°C. It solidifies at low temperature to glacial acetic acid. It is used for making vinegar.

Acetoin. 3-hydroxy-2-butanone. $CH_3CH(OH)COCH_2$. It is a yellow *liquid*, b.p. 148°C. It is used in the manufacture of flavours.

Acetolysis. The process of conversion of a group of atoms in an organic compound to ethanol (acetyl). It is carried out by reacting the compound with *glacial ethanoic acid.*

Acetone (CH_3COCH_3). Colourless, inflammable liquid. It has pleasant odour and boils at 56.5°C. It is used as solvent in production of cellulose acetate rayon.

Acetone Pyrolysis. Thermal decomposition of acetone to give ketene.

Acetonitrile. Ethanenitrile, methyl cyanide CH_3CN. It is a colourless poisonous *liquid* b.p. 82°C. It smells like *ethoxyethane* (ether). It is used in organic synthesis and as a *solvent.*

Acetylase. An enzyme that catalyses formation of acetyl esters.

Acetylen (C_2H_2). Colourless, poisonous, inflammable gas.

Acid. Sharp, corrosive substance having sour taste; turns litmus red; liberates hydrogen ions in solution and reacts with a base to form a salt.

Acid Dyes. Refer to a group of *dyes* ; these are used chiefly for dyeing wool and natural silk from an acid dyebath.

Acid Halide. Acyl halide. An organic compound with the general formula RCOX. They can be obtained from carboxylic acids by replacing the hydroxyl group with a halogen atom. They are used in *halogenation.*

Acidic Hydrogen. That hydrogen atom in an *acid* that is replaceable by *metals* to form *salts.*

Acidimetry. The process of determination of the amount of *acid* present in a *solution* by *titration.*

Acidolysis. *Hydrolysis* using an *acid.*

Acid radical. A *molecule* of an *acid* without the ***acidic hydrogen, e.g.,*** the SO_{42-}.

Acid Rain. Rain having acidic pH. It is due to presence of oxides of nitrogren and sulphur in atmosphere.

Acid Salt. An *acid* in which only a part of the *acid hydrogen* has been replaced by a *metal e.g. sodium bicarbonate,* $NaHCO_3$.

Acid Value. It is a measure of the free *fatty acid* present in a fat or oil. It is equal to number of *milligrams* of *potassium hydroxide* required to completely neutralize the free fatty acids in one gram of the substance.

Acoustics. 1. The study of *sound.* 2. The characteristics of an auditorium that determines its ability to enable music and speech to be heard clearly within it.

Acoustic Spectrum. The range of frequencies that occur in the sound emitted by a source.

Acoustoelectronics. The study and use of devices in which *electronics signals* get converted by *transducers* into surface acoustic waves and passed through tiny solid strips. When acoustic signals are propagated some 10^5 times more slowly than *etectromagnetic waves,* this technique enables *delay lines* to be constructed that can be up to 50 times lighter than pure electronic devices.

Acre. Unit of area, one acre is equal to 4840 square yards or 4046.86 square metres.

Acriflavine. 3,6-Diamino-10-methylacridinium chloride. $C_{14}H_{14}$ N_3Cl. A yellow substance used as an *antiseptic.*

Acrolein. Colourless, pungent smelling liquid. It is an aldehyde of allyl alcohol.

Acrylic Resins. *Plastics* obtained by the *polymerization* of derivatives of *propenoic* (acrylic) *acid.*

ACTH. It is a *polypeptide hormone* secreted by the pituitary gland. It controls the secretion of corticosteroid hormones by the adrenal glands.

Actinic Radiation. *Electromagnetic radiation* they can cause photochemical reactions. These radiation can be used as a source of illumination in photography.

Actinides. Actinoids, actinons. The *elements* with *atomic numbers* from 89 (actinium) to 103.

Actinium. Ac. Element. R. a.m. 227. At. No. 89. A *radioactive* substance, *half-life* 21.6 years m.p. 1050°C., b.p. 3200°C. There are two natural isotopes. Ac. 227 (0.7% occurs in natural uranium) and Ac. 228. Some 20 artificial isotopes are known.

Actinometer. It is an instrument used to measure heat intensity.

Actinometer. An instrument used to measure the intensity of *electromagnetic radiation.* It is based on fluorescence or a photographic process.

Actinomycosus. Lumpy-jaw (disease) in cattle.

Actinothrapy. Disease treatment by exposure to ultra-violet rays.

Acyclic. Having an *open-chain* structure.

Acyl. The *univalent radical* RCO—, where R is an organic group.

Acylation. The process of introduction of an *acyl group*, RCO into a compound.

Adatom. An adsorbed *atom.*

Adaxial. Upper surface of leaf.

Adder. Small poisonous snake.

Adductor. Muscle that draws part of body towards median axis.

Addition Compound. A chemical *compound* formed by the addition of an *atom* or group of atoms to a *molecule e.g.*, addition compound of *carbon monoxide*, CO, and chlorine, Cl_2 is phosgene $COCl_2$.

Addition Reaction. A *chemical reation* in which *double bond* in an *unsaturated compound* is converted to a single bond by the addition of other *atoms* or groups.

Additive Process. The process of *colour* formation by a mixture of red, green, and blue lights. These colours add together to form a new colour.

Adduct. An *addition compound* formed by a reaction involving no change in valency.

Adenine. 6-Aminopurine, $C_5H_3N_4NH_2$. A white crystalline *purine* base, m.p. 360-365°C. It occur in-*nucleic acids* and in *adenosine tripho- sphate.*

Adenosine. A *nucleoside* consisting of *adenine* linked to a D-*ribose sugar* molecule.

Adenosine Triphosphate. ATP. $C_{10}H_{12}N_5O_3H_4P_3O_9$. A *nucleotide* which helps transfer of *energy* within living *cells.*

Adermine. Vitamin essential for growth of lactic acid bacteria, certain fungi and yeasts.

Adhesives. Substances used for sticking surfaces together ; *e.g., glues cements*, etc.

Adiabatic. Refers to changes taking place without *heat* entering or leaving the system.

Adiabatic Demagnetization. A method used for attaining temperatures in the region of *absolute zero* by magnetizing a *paramagnetic* substance such as potassium chrome alum or gadolinium sulphate, and allowing it to demagnetize adiabatically.

Admittance. *Y.*, It is the reciprocal of *impedance.*

Adipic Acid ($C_6H_{10}O_4$). Acid obtained by action of fat with nitric acid.

Adipocere. Fatty substance obtained as resulting of decomposition of animal bodies in moist places or under water.

Adipose. Refers to Fatty.

Adipose Tissue. Vasicular structure in animal body containing fat.

Admittance Y. It is the reciprocal of impedence.

Adrenaline Epinephrine. 3, 4-Dihydroxy-α-(methylaminomethyl) benzyl alcohol, $C_9H_{13}NH_3$. It is a *hormone* produced by the *medulla* of the adrenal glands and synthetically. It acts as a heart stimulant. It constricts blood vessels.

Adsorbate. The substance that is absorbed.

Adsorbent. A substance that adsorbs.

Adsorption. It is a surface process *e.g., molecules* of a *gas* or of a dissolved or suspended substance come on the surface of a *solid.*

Advection. The process of transfer of either *matter* or *energy* from one place to another by a horizontal stream of *gas.*

Aerenchyma. Refers to respiratory tissue found in hydrophytes and semi-aquatic plants.

Aerial. It is the part of the radio system from which the energy is transmitted to or received from space.

Aerobe. These organism that require air or oxygen so as to maintain the life process.

Aerobic. In presence of oxygen in free state.

Aerodynamics. The study of the motion and control of solid bodies in *air*. The study of air or other *gases* in motion.

Aerogenerator. A device to extract usable energy from winds.

Aerolites. *Meteorites*, especially those which consist of stony material rather than iron.

Aero Metal. An alloy consisting chiefly of aluminium, zinc, and copper.

Aerosol. A dispersion of *solid* or *liquid* particles in a *gas e.g. smoke.*

Aerospace. It is the *Earth's atmosphere* and the *space* beyond.

Aetiology. (U.S.A., etiology) The science of causation. It is used in medicine where it means the science of the causes of disease.

Affinity. The energy that binds the atoms together.

Affinity (chem.). Chemical attraction; the *energy* that binds *atoms* together.

Affatoxins. Refers to four related toxic compounds produced by the mould *Aspergillus flavus*. These occur in peanuts and cereals that are contaminated with the mould and they can cause liver damage and cancer.

Affine. A affine transformation of the plane is one that can be represented in the form

$$\begin{pmatrix} x \\ y \end{pmatrix} \rightarrow \begin{pmatrix} a & b \\ c & d \end{pmatrix} \begin{pmatrix} x \\ y \end{pmatrix} + \begin{pmatrix} e \\ f \end{pmatrix}$$

Examples include translations, rotations, shears, etc.

Afterburning 1. The *combustion* that results from the addition of *fuel* to the exhaust of a *jet engine.* It increases *thrust* and reduces fuel consumption. **2.** The irregular burning of residual *propellant* in a *rocket* motor when the main combustion is over.

After-Damp. A poisonous mixture of *gases.* It contains *carbon monoxide* formed by the explosion of *fire-damp* in coal-mines.

After-Glow. Glow that is sometimes observed high in the western sky after sunset. It is caused by fine dust particles present in the *upper atmosphere* that scatter the *light* from the *Sun.*

After-Heat. *Heat* generated in a *nuclear reactor* when it has been shut down.

Agamic. Refers to Reproduction without union of gametes.

Agar Agar. It is gelatinous material obtained from certain sea weeds; It is used as base for culture media and also for setting jellies and puddings.

Agate. Precious stone, very hard natural form of silica. It is composed of layers of quartzs of different colours; It is often used in ornaments.

Aglycone. A non-sugar component present in glycoside.

Agonic line. It is a line which has zero magnetic declination.

AGR. Used for advanced gas cooled reactors.

Air. Gaseous mixture that surround the earth; in addition to atmospheric gases it contains water vapour, hydrogen, hydrogen peroxide, sulphur compounds and also very small dust particles.

Air Equivalent. The thickness of a layer of air at *S.T.P.* which causes the same amount of absorption of nuclear radiation as the substance being considered.

Alabaster. A natural opaque form of ***hydrated calcium sulphate***, $CaSO_4 . 2H_2O$.

Alanine. A colourless crystalline *amino* acid soluble in water.

Albedo 1. It is the ratio of the *radiant flux* reflected by a surface to that falling on it. **2.** The probability that a *neutron* entering a material will be reflected back by that material through the surface by which it entered.

Albumins Albumens. A group of *soluble* globular *proteins* these occur in many animal *tissues* and *fluids*; e.g. egg-white (egg-albumin), milk (lactalbumin), and *blood* (serum albumin).

Alchemist. Well versed in Alchemy.

Alchemy. An art to transmute base *metals* into gold. It flourished from about A.D. 500 till the Middle Ages, when it gradually fell into disrepute.

Alcohalates. Metallic *salts* of *alcohols*. These are formed by replacement of hydrogen atoms in the *hydroxyl* groups of the alcohol by metals, e.g., sodium ethanolate (sodium ethoxide), C_2H_5ONa.

Alcohometry. The process of determination of the proportion of *ethanol* in spirits and other *solutions*; it is usually performed by measuring the relative density of the liquid at a standard *temperature* by a specially graduated *hydrometer.*

Alcohols. A class of *organic compounds*. The names of alcohols are obtained from the hydrocarbons from which they are derived to which the suffix *-ol* is added. e.g. *ethanol* (ordinary 'alcohol') is C_2H_5OH, theoretically derived from *ethane*, C_2H_6.

Alcohol (C_5H_2OH). Theoretically derived from ethane (C_2H_6). It is distillate obtained by fermentation of sugar or other saccharine matter, constituting intoxicating element in liquors.

Aldicarb ($C_7H_{14}N_2O_2S$). Crystalline, white solid having slightly sulphurous smell. It is soluble in water but readily soluble in organic solvents. It is used as systematic insecticide, acaricide and nematicide.

Aldehyde. A class of organic compounds with general formula RCHO. They are named on IUPAC system by changing of parent hydrocarbon by-*al.*

Aldose. It is a mono saccharide having a formyl group in the molecule.

Aldrin. ($C_{12}H_8Cl_2$)—used as soil insecticide mainly against termites.

Algebra. The generalisation and representation in symbolic form of significant results and patterns in arithmetic and other areas of mathematics.

Algebraic Sum. Total of quantities of the same type with due regard to sign.

Algin. A loose term used for *alginic acid* or its sodium salt.

Alginic Acid. $(C_6H_8O_6)n$. A complex *organic compound.* It is used for preparing *emulsions* and as a thickening agent in the food industry.

Algol. Algorithmic language. A type of *computer* language. It is based on *Boolean algebra,* for expressing information in an algebraic notation.

Algoly. It is the scientific study of algae.

Algorithm. Algorism (math.) A systematic mathematical procedure which enables a problem to be solved in a finite number of steps.

Alicyclic Compound. A type of *aliphatic,* organic compound that contains a *saturated* ring of carbon atoms.

Alidade. An instrument used for measuring vertical heights and distances.

Aliphatic Compounds. *Organic compounds* containing open chains of carbon *atoms.* They consist of the *alkanes, alkenes,* and *alkynes* as well as all their *derivatives* and *substitution products.*

Aliquot Part. A divisor of a number or quantity that will give an *integer.* Thus 3 is an aliquot part of 6, but 5 is not.

Alizarin. 1, 2-Dihydroxyanthraquinone. $C_{14}H_6O_2\ (OH)_2$. An orange-red crystalline substance m.p. 289°C. Used in dyeing with the aid *mordants.*

Alkali. A *soluble hydroxide* of a *metal,* particularly of one of the *alkali metals.*

Alkali Metals. The univalent metals lithium, sodium, potassium, rubidium and caesium. They belong to Group 1A of the *periodic table.*

Alkalimetry. The process of determination of the amount of *alkali* present in a *solution,* by *titration.*

Alkaline. Having the properties of an *alkali.*

Alkaline Earth Metals. The *bivalent* group of *metals* comprising beryllium, magnesium, calcium, strontium, barium, and radium. They belong to Group 2A of the *periodic table.*

Alkaloid. Nitrogenous, basic organic compound. It is found in some dicotyledons plants.

Allobar. Mixture of isotopes of an element which do not occur naturally.

Allotropes. Different physical forms of solid substances *e.g.*, diamond graphite are allotropes of carbon.

Allotropy. Property of a substance of existing in two or more different forms.

Alloy. An alloy is a mixture which is made up of two or more metals or which contains metals and non-metals. Properties of an alloy differ from the constituents.

Alpha (α). First letter of Greek alphabet.

Alpha decay. Radioactive decay in which nucleus spontaneously emits alpha particles.

Alpha-Iron. An allotropic form of pure iron. It exists up to 900°C.

Alpha particle. A helium *nucleus*. Alpha particles are emitted from the *nuclei* of certain *radioactive elements.*

Alpha Radiation. They consist of streams of fast-moving *alpha particles.* Alpha radiation produces intense *ionization* in *gases* through which it passes. It is easily absorbed by *matter*, and produces *fluorescence* on a fluorescent screen.

Altazimuth. An instrument used for the measurement of the *altitude* and *azimuth* of heavenly bodies.

Alternating Current *a.c.* A flow of *electric current* that, after reaching a maximum in one direction, decreases, finally reversing and reaching a maximum in the opposite direction, the *cycle* being repeated continuously.

Alternator. A machine that produces electrical *alternating currents.*

Altimeter. It is an instrument used to measure height above sea-level. It usually consists of an *aneroid barometer* calibrated to read zero at sea-level and the height above sea-level in metres or feet.

Altitude. 1. Height. **2.** The altitude of a heavenly body is its *angular distance* from the horizoa on the vertical circle passing through the body, the *zenith* and the *nadir.*

Alum. Potash alum. $K_2SO_4.\ Al_2(SO_4)_3.\ 24H_2O$. Crystalline aluminium potassium sulphate. The ***compound*** occurs naturally and is used as a *mordant* in dyeing for fireproofing, and other technical purposes.

Aluminium Paint. Prepared by mixing aluminium pigment in varnish; it reflects sun's radiation and retains heat in hot-air or hot water pipes and tanks.

Aluminium Paste. Aluminium powder finely ground in oil ; used in aluminium paints.

Alumina. *Aluminium oxide.*

Aluminium Silicates. Minerals having aluminium and silicon combined with oxygen in their structure *e.g.*, clay, micas, zeolites etc.

Aluminium Soap. Salt of higher carboxylic acid and aluminium. It is insoluble in water but soluble in oils; used in greases paints, varnishes etc.

Alunite. Natural compound of potassium, aluminium sulphate and aluminium hydroxide. It is also known as alum-stone. It is a source of alum.

Alveolus. Refers to small cavity or pit. Tooth socket. Air-sac in lung.

Amalgam. Alloy or mixture of mercury with any other metal.

Amalgamation Process. Used for metallurgy of gold. Gold-bearing rock or sand, after crushing, is treated with mercury, which forms an *amalgam* on the surface of the gold. The amalgamated particles are allowed to stick to amalgamated copper plates, the rest of the ore being washed away; they are then removed.

Amatol. An explosive mixture contains about 80% NH_4NO_3 and 20% T.N.T.

Amber. A fossil resin obtained from exinct species of pinc. It contains succinic acid and is used for ornamental purposes. It is also used as an electric insulator.

Ambergris. A black waxy substance found in the intestine of Sperm whale. It is used in perfume.

Ammeter. Instrument used for measuring electric current.

Ammonal. Mixture of ammonium nitrate, powdered aluminium and tri-nitrotoluene (TNT); used as explosive.

Ammonia (NH_3). A gas that has pungent smell very soluble in water, giving alkali solution containing ammonium hydroxide (NH_4OH); used in manufacture of explosives, fertilizers etc.

Ammoniated. Impregnated with ammonia.

Ammonifying Bacteria. Bacteria which decomposes protein and other nitrogenous substances to form ammonia.

Ammonium Acetate (CH_3COONH_4). White deliquescent solid; It is used in manufacture of dyes and as metal preservative.

Ammonium Bicarbonate (NH_4HCO_3). White crystalline solid; It is used as baking powder.

Ammonia Clock. An atomic clock which is based on the vibrational frequency with which the nitrogen atom in the *ammonia* molecule passes through the plane of the three hydrogen atoms and back again. The vibration possess a frequency of 23870 *hertz* and a *quartz crystal* is used to supply ammonia gas with energy at this frequency. As ammonia will only absorb energy at this frequency, the ammonia can be used to regulate the frequency of the quartz oscillator, through a feedback circuit.

Ammonium Carbonate ($(NH_4)_2CO_3$). A white *soluble* crystalline salt, usually occurs as the monohydrate $(NH_4)_2CO_3 . H_2O$ decomposes slowly, forming ammonia, carbon dioxide and water.

Ammonium Chloride. Sal ammoniac. NH_4Cl. A white crystalline *salt*, used in *dry cells* and *Leclanche cells*.

Ammonium Ethanoate. Ammonium acetate. CH_3COONH_4. A white solid. m.p. 114°C., used as a meat preservative and in the manufacture of *dyes*.

Ammonium Hydroxide NH_4OH. A *compound* exists in *aqueous solutions* of *ammonia*; the name is often applied to the *solution*.

Ammonium Nitrate. NH_4NO_3. A white crystalline *salt*, m.p. 169.6°C. It decomposes on heating to form *dinitrogen oxide*, N_2O, and water; used in *explosives*, e.g., *ammonal, amatol*.

Ammonium Ion NH_4^+. A *univalent* ion that in *compounds* behaves like an *alkali metal*, giving rise to ammonia *salts*.

Ammonium Sodium Hydrogen Orthophosphate. Microcosmic salt. $NH_4NaHPO_4 . 4H_2O$. A white crystalline *salts*, used as a *flux*.

Ammonium Sulphate ($(NH_4)_2SO_4$). A white crystalline *salt*, obtained as a by-product in manufacture of coal gas produced from *ammonia* and *sulphuric acid*, used as a *fertilizer*.

Ammonium Thiocyanate NH_4SCN. A colourless crystalline substance, m.p. 149.6°C., used as a herbicide and in the textile industry.

Amorphous. Non-crystalline, having no definite form or shape.

Ampere. It is a unit of electric current; approximately equal to flow of 6×10^{18} electrons per second, when passed through solution of silver nitrate, it will deposit 0.001118 gram per second of silver.

Ampere-Hour. The practical unit of *electric charge*, the charge flowing per hour through a *conductor* when the current in it is one *ampere*. 3600 *coulombs*.

Ampere's Law. The strength of the *magnetic field* induced by a *current* flowing through a *conductor* is, at any point, directly proportional to the *product* of the current and the length of the conductor and inversely proportional to the *square* of the distance between the point and the conductor. The direction of the field is perpendicular to the *plane* joining the point and the conductor.

Ampere-Turns. A unit of *magnetomotive force*. It is equal to the product of the number of turns in a coil and the current in *amperes* that flows through it.

Amphetamine. $C_6H_5CH_2CHNH_2CH_3$. A *drug*, used in the form of the *sulphate*, stimulates the central nervous system in cases of depression. Also know under the trade name, Benzedrine.

Amphiboles. Complex silicates including *hornblende* and *asbestos*.

Amphichroic. Amphichromatic. Produces one colour on reaction with an *acid* and another colour on reaction with a *base*.

Amphiprotic. Capable both of accepting and of giving any *protons* in solution; *amphoteric*.

Ampholyte. An *amphoteric* substance, that is an *electrolyte*.

Amphoteric. Behaves as *acidic* to strong *bases* and as *basic* towards strong *acids*, *e.g.*, the amphoteric *oxide*, *zinc oxide*, gives zinc *salts* with strong *acids* and zincates of the *alkali metals*.

Amplifier. An *electronic* device. It increases the strength of a signal fed into it, by obtaining *power* from a source other than the input signal.

Amplitude (phys.) If any quantity is varying in an oscillatory manner about an equilibrium value, the maximum departure from that equilibrium value is called the amplitude; *e.g.*, in the case of a *pendulum* the amplitude is half the length of the swing.

Amplitude Modulation. A.M. One of the methods to transmit information by *radio* waves. The *amplitude* of a *carrier wave* is modulated in accordance with the *frequency* of the signal to be transmitted.

Amylose. Component (15 to 20%) of most cereal starches, soluble in water; polysaccharide whose molecules consist of long unbranched chain of glucose units.

Amylum. Starch.

Anaerobe. Organism that does not require free oxygen for living.

Anemophily. Air pollination.

Aner. Male insect especially ant.

Anesthesia. Total or partial loss or sensitivity particularly to touch.

Anethole. $CH_3CH{=}CHC_6H_4OCH_3$. A white crystalline powder used in perfumes, flavouring and medicines.

Angle. The space between two intersecting lines or *planes*; measured in *degrees* or in *radians.*

Angstrom Unit. A.U., PO^{-10} *metre.* A unit of length, used or measurement of *wavelength* of *light* and intra-molecular distances. 10 A.U.=1 nanometre.

Angular Acceleration. It is the rate of change of *angular velocity.*

Angular Displacement. The *angle* through which a point, line or body has been rotated in a specified directions, about a specified *axis.*

Angular Distance. The distance between two bodies, measured in terms of the *angle* subtended by them at the point of observation, used in *astronomy.*

Angular Frequency. The *frequency* of a periodic process expressed in *radians* per *seconds*; it is equal to 2π times the number of cycles per second.

Angular Momentum L. The product of *moment of inertia* and *angular velocity.*

Angular Velocity ω. Rate of motion through an *angle* about an *axis*. It is measured in *degrees, radians*, or revolutions per unit time.

Anhydride. The anhydride of a substance is that which, when chemically combined with *water*, gives the substance.

Anhydrite. A naturally occurring form of *calcium sulphate*, $CaSO_4$.

Anhydrous. Having no water, it is often applied to *salts* without *water of crystallization.*

Anilide. An *organic compound* similar to an *amide*, derived from an *aromatic amine*, especially from *phenylamine* (aniline).

Aniline. ($C_6H_5NH_2$). Colourless, poisonous oily liquid, has unpleasant odour, insoluble in water, boiling point 185°C. It is used in manufacture of dyes, drugs, plastics etc.

Aniline Dye(s). The Dyes which are prepared or chemically derived from aniline.

Aniline Ink. Fast drying printing ink. It is a solution of coal-tar dye in organic solvent.

Animal Charcoal. Material containing carbon (10%) and inorganic matter mainly calcium phosphate (90%); It is used as decolourising agent.

Anion. Electro-negative ion, attracted towards anode.

Anisaldehyde. Colourless oily liquid. It is used in cosmetics and perfumes.

Anisole. Methyle phenyl ether ($CH_3OC_6H_5$). It is a colourless liquid with aromatic odour. It is used in perfumes.

Anisomerous. Having unequal number or parts in floral whorls.

Anisotropic. Not isotropic ; showing different physical properties in different directions.

Annual Variation. A very small but regular variation that the *magnetic declination* undergoes in the course of a year.

Annular. An annular space is the space between an inner and outer ring.

Annular Eclipse. An *eclipse* of the *Sun* in which a ring of its surface is visible surrounding the darkened *Moon.*

Annulus. A plane figure consisting of the area between two concentric circles of different radii. Its area is $\pi(R^2-r^2)$, where R and r are the two radii.

Anode. Positive electrode by which electric current enters electrolyte.

Anopheles. Genus of germ carrying mosquitoes.

Anoplura(Siphonaptera). Order of wingless blood sucking insects (1400 speçies); mostly ectoparasites of vertebrates.

Antacid. A drug that counteracts stomach acidity *e.g.* Al $(OH)_3$ Mg $(OH)_2$ etc.

Anthocyanin. Glucoside plant pigment. It is violet in neutral, red in acid and blue in alkaline.

Anthracene ($C_{14}H_{10}$). White, crystallinc polycyclie hydrocarbon. It has slight bluish flourescence, boiling point 355°C; It is obtained from coal-tar by fractional distillation; It is a source of dyestuffs.

Anthracite. Stone-coal consisting almost of carbon. It burns neatly without flame or smoke.

Antiauxin. Chemical which prevents action of auxins in plants.

Antimoniats. These are salt of any antimonic acid.

Antimonic. Salts containing pentavalent antimony.

Antimonious. Salts containing trivalent antimony.

Aantimonite. Salt of antimonious acid.

Antimony (Stibium, Sb). Brittle crystalline silvery-white metal. It is a poor conductor of heat and electricity, melting point 360°C. It is used in type metal and other alloys to which it adds hardness and resistance to corrosion.

Antimony Pentachloride ($SbCl_5$). Colourless liquid obtained by treating molten trichloride with chlorine; used as strong chlorinating agent.

Aantimony Pentasulphide (Sb_2S_5). Yellow powder insoluble in water; It is used in vulcanisation of rubber.

Antimony Sulphate [$Sb_2(SO_4)_3$]. White crystalline insoluble solid ; It is used in explosives.

Antimony Trichloride ($SbGl_3$). White soft solid, very deliquescent; used as caustic in medicines, for cleaning gun barrels.

Antioxidant. Substance used to prevent harmful effects of oxidation in paints, plastic, rubber etc.

Antipyretic. Drug used for lowering temperature in case of fever.

Antisepsis. Destruction of bacteria.

Antiseptic. Any substance used for destroying or stopping growth of micro-organisms in living tissues.

Antitoxin. Any chemical that can neutralise poison or toxin.

Apatite [$Ca_5\ (PO_4)_3F$]. Natural phosphatic mineral having fluoride and phosphate anions; used for production of phosphorus and its compounds and in manufacture of fertilizers.

Aperture. Size of opening, that admits light in optical instruments like camera, microscope etc.

Apocynthion. Refers to the time or point, in orbit of a lunar satellite when it is farthest from the sun.

Apoenzyme. Protein part of enzyme.

Apomorphine ($C_{17}H_{17}NO_2$). Crystalline alkaloid. It is derived from morphine.

Apozymase Protein part of zymase.

Aprotic Solvent. A solvent that does not yield or accept a proton.

Aqua. Water.

Aqua-Fortis (HNO_3). Concentrated nitric acid.

Aqua-ion (s). Hydrated metal ions in aqueous solution.

Aqua-Marine. Sea-water.

Aqua-Mirabilis. Refers to preparation distilled from cloves, ginger, nutmeg and spirit of wine.

Aqua-regia. A solution of hydrochloric acid and nitric acid (3 : 1). It is highly corrosive liquid and can dissolve noble metals (gold and platinum).

Aquarium (pl. aquaria). Tank or series of tanks in which aquatic plants or animals are kept.

Aquarius. Water-bearer sign of Zodiac.

Aquatint. Etching on copper with resin and nitric acid.

Aqua-vitae. Liqueur—brandy, Whiskey etc.

Aqueous. Watery.

Aqueous Humour. Watery fluid between cornea and lens in the eye.

Arabinose $C_5H_{19}O_5$. White soluble crystalline pentose sugar. It is obtained from gums or synthetically from glucose. It is used as culture medium in bacteriology.

Araeometer. An instrument (hydrometer) used for measuring specific gravity.

Archimedes' Principle. It states that the loss in weight of a body fully or partially immersed in liquid is equal to weight of liquid displaced.

Arene. Aromatic hydrocarbon.

Argenite (Ag_2S). Silver sulphide which occurs in nature.

Argentometer. An Instrument (hydrometer) used to detect amount of silver salt in a solution.

Arginine. An essential amino acid.

Argentometry. Volumetric analysis that employs precipitation of insoluble silver salts.

Argil. It refers to potter's clay.

Argol. Tartar, Reddish brown deposit which separates in wine vats. It consists mainly of pot., hydrogen tartarate.

Argon. Ar. Element. R. a.m. 39.948. At. No. 18 An *inert gas*, m.p.—189°C., b.p.—185°C., present in the air (0.9%). Used for filling electric lamps and in fluorescent tubes at a pressure of about 3 mm of mercury.

Argument (math.)1. An independent *variable* that forms part of a *function.*

Arithmetic Mean. Arithmetic average. The sum of a set of *n* numbers divided by *n. e.g.*, the arithmetic mean of 4, 3 and 8 is 5.

Arithmetical Series. A *series* of quantities in which each term differs from the preceding by a constant common difference.

Armature. Coils of dynamo of electric motor. It is that part of dynamo which rotates in magnetic field and develops electric current *e.g.*, gramophone pick-ups, loud speakers etc.

Armilla. Frill on mushroom stalk.

Arolium. Median pad between claws of insect's foot.

Aroma. Spicy fragrance.

Aromatic Compounds (s). Class of organic compounds containing at least one benzene ring.

Arsenate. They are Salt of arsenic acid.

Arsenic (As). An element, available in three crystalline allotropic forms *i.e.* α-arsenic (yellow), β-arsenic (black) and γ-arsenic (grey) — compounds of arsenic are highly poisonous. They are used in preparing various pesticides etc.

Arsenic Oxide (As_2O_5). White solid readily soluble in water to form a solution of arsenic acid, $As_2O_5+H_2O \rightarrow H_3AsO_4$; It can be obtained by heating arsenious oxide with oxygen at 300°C under a high pressure. It is used as an oxidising agent.

Arsenious Oxide (As_2O_3). Glassy amorphous powder which soon becomes opaque and reverts to crystalline form; used for making free arsenic and in insecticides and herbicides.

Arsine (AsH_3). Colourless gas with garlic-smell, insoluble in water. It is prepared in laboratory by action of metal and dilute acids on arsenious oxide; It burns in air with bluish-white flame, forms water and arsenious oxide and used for detection of arsenic.

Arsonophrites. Mineral compound containing arsenic, iron and sulphate.

Artery. Blood vessel that carries blood away from heart; branch of aorta that further divides into arteioles.

Arthromere. Body segment of an articulated animal; somite.

Arthroscope. An instrument used for visualisation of anterior of joint cavity.

Articulation. It is the Percentage of speech-sounds correctly received over adio-communication or other reproducing systems.

Artificial Gold (SnS_2). Yellowish-brown powder, insoluble in water. It dissolves in alkaline sulphides and is used for imitation gilding etc.

Artificial Sunlight. Light obtained from lamps rich in ultraviolet rays.

Aryl. Organic univalent radical derived from an arene.

Asbestos. Fibrous silicate mineral contains mainly calcium, magnesium silicate and is used as a heat insulating material.

Ascorbic Acid. Vitamin $C_6H_8O_6$. Deficiency causes scurvy.

Ash. Incombustible residue obtained after complete combustion of any substance.

Asphalt. Dark brown to black, semi-solid sticky substance, made up of bitumen with mineral matter, occurs naturally in asphalt lakes obtained as residue in petroleum distillation; It is used as cement in buildings and in paints, varnishes etc.

Astatine (At). Radioactive element. It is more metallic than iodine, boiling point 337°C. It can be obtained by bombarding bismuth nuclei with alpha particles.

Asteroids. A belt of small bodies rotating round the *Sun* in orbits between those of *Mars* and *Jupiter.*

Astigmatism. A defect of *lenses* caused due to the curvature being different in two mutually perpendicular *planes*; thus *rays* in one plane may be in focus while those in the other plane are out of focus, producing distortion. Astigmatism of the eye can be corrected by the use of cylindrical lenses.

Astringent. A substance that reduces the discharge of mucus or *blood* by contracting body tissues.

Astrocompass. An instrument used for determining direction relative to the *stars*. It is used to determine the errors of such instruments as magnetic or gyro compasses.

Astrolabe. An instrument used by early astronomers to measure the *altitude* to heavenly bodies.

Astrology. The pseudo-science of predicting the course of human destinies by indications based on positions and movements of the heavenly bodies.

Astrometry. The branch of *astronomy* concerned with measurements of the positions of celestial bodies on the *celestial sphere.*

Astronautics. The scientific study of travel outside the *Earth's atmosphere.*

Astronomical Unit. It is the mean distance from the centre of the *Earth* to the centre of the *Sun*. 1.495×10^{11} *metres*, approximately 92.9×10^{6} miles.

Astronomy. The scientific study of the heavenly bodies, their motion, relative positions and nature. Various branches are *astrometry, celestial mechanics* and *astrophysics.*

Astrophysics. The branch of *astronomy* concerned with the physical properties of celestial bodies, and the interaction between *matter* and *energy* within them.

Asymmetric. Not possessing *symmetry.*

Asymmetric Carbon Atom. A carbon *atom* in a *molecule* of an *organic compound* with four different atoms or groups attached to its four *valences*. Such a grouping leads to the existence of optical isomers.

Asymptote. A line approaching a curve, but never reaching it within a finite distance.

Atactic Polymer. A *polymer* in which the groups attached to the main chain are not arranged regularly. In isotactic polymers the same irregularity is repeated along the chain, whereas in syndiotactic polymers there are *asymmetric carbon atoms* in the chain and successive groups lie on alternate sides of the chain.

Athermancy. Property of being opaque to radiant heat.

Atmosphere. Gaseous layer surrounding earth ; its composition is nitrogen (78%), oxygen (21%), argon (0.95%), carbon dioxide (0.04), neon (0.002%), helium (0.0005%), krypton (0.0001%), xenon (0.00001%) and some other gases in still small measure.

Atmospheric Noise. Noise heard during radio reception on account of atmospheric interference.

Atmospherics. Electrical discharges that take place in atmosphere. They cause crankling sound in radio receivers.

Atoll. Coral island consisting of horse-shoe-shaped or circular belt of coral reef surrounding central lagoon.

Atom. Smallest unit of element that cannot be further sub-divided and is capable of taking part in chemical reaction.

Atom Bomb. Nuclear weapon. It is capable of causing wide spread destruction, explosion is caused by fast uncontrolled nuclear fission reaction.

Atomic Clock. Type of very accurate clock which is based on vibrations of atoms and molecules.

Atomic Energy. Nuclear energy.

Atomic Heat. Product of atomic weight and specific heat of an element.

Atomic Number. Number of electrons or protons present in a neutral atoms of an element ; also called proton number.

Atomic Pile. Early form of nuclear reactor constructed by piling up graphite rocks and uranium rods.

Atomic Theory. Hypothesis that all atoms of same element are alike.

Atomic Volume. Atomic weight of element divided by its density.

Atomic Weight. Weight of an atom relative to that of hydrogen as I and (now) oxygen as 16.

Atomize. To break into atoms.

Atomizer. Very small sprayer for discharging liquid in very fine spray.

Atom Smasher. A Particle accelerator.

Atrophied. Refers to rudimentary or vestigial.

Atropine ($C_{17}H_{23}NO_3$). Colourless, crystalline, poisonous, insoluble alkaloid; used in medicines to dilate pupil of eye.

Attenuation. Reduction in density, force or intensity of radiation when it passes through matter.

Attractant. Chemical that lures insects.

Audio. Frequencies corresponding to sound waves that are normally audible. Relates to sound reproduction.

Audio-frequency. Frequency of oscillation that lies in range of 30 to 2000 hertz; it is perceptible by human ear.

Audiometer. An instrument for measuring human hearing.

Audiotypist. Typist trained to type from dictating machine.

Adiovisual. Involving both sight and sound; used as teaching aid.

Auric (Au). Trivalent gold. Brown metal, deliquescent and readily soluble in water; used in electroplating.

Auriferous. Gold-bearing.

Aurous (Au). Univalent gold, yellow coloured transition metal. It is used for making ornaments and coins.

Aurous ion (Au^+). Powerful oxidizing agent oxidizes water. It does not exist as simple cation in aqueous solution.

Autocatalysis. Reaction or disintegration of cell or tissue due to influence of one of its own product.

Autoclave. Strong vessel in which super-heated steam under pressure is used for sterilising or cooking.

Autogrio (autogyro). A type of aircraft having both propeller and a large horizontal rotor.

Autolysis. Self-destruction of biological cells after death.

Autonomic. Part of nervous system that functions independent of brain.

Autophagous. Self-devouring ; Refers to some birds which eat or bite off part of their own body.

Auxanometer. Instrument for measuring linear growth of plant parts.

Auxin. Phytohormone; growth promoting substance found in minute quantities in higher plants.

Avalanche. Refers to shower of particles caused by collision of high energy particle with any other form of matter.

Avitaminosis. Disease due to lack of vitamins.

Avogadro's Law. Equal volumes of all gases under similar conditions of temperature and pressure contain same number of molecules.

Azeotrope. Mixture of two or more liquids which distill without change in their composition *e.g.*, solution of hydrochloric acid in water.

Azetidine. Colourless liquid. It smells like ammonia and boils at 61°C.

Azide. Compound containing ion N_3 — or univalent azido group N_3— potassium azide (KN_3), sodium azide (NaN_3), lead azide [$Pb(N_3)_2$] etc.

Azine. Organic derivative of hydrazine containing C=N—N=C group.

Azino. The *quadrivalent* radical =N.N=

Azo Compound. A compound containing an *azo group* attached to two carbon atom (– CN=NC–). *Aromatic* azo compounds are usually prepared by *azo coupling.*

Azo Coupling. The process of formation of an *azo compound* by the reaction of an *aromatic diazo compound* with a suitable *nucleophilic reagent*, such as an *amine* or a *phenol.*

Azo Dyes. *Azo compounds* used as dyes.

Azoic Dyes. Insoluble *azo dyes* are those which are formed within the fibre by the *azo coupling* of a *diazo compound* and a suitable *azo-coupling* component, generally a *naphthol* derivative.

Azote. Former name for *nitrogen.*

Azurite. Naturally occuring *basic* copper (II) carbonate, $2CuCO_3 . Cu(OH)_2$. It has an intense blue colour. It is used as a gemstone.

B

Babbit Metal. An alloy contains high proportion of tin or lead and small amounts of antimony and copper; It is used for making bearings.

Babo's Law. When a non-volatile solid is added to a liquid, in which it is soluble, it lowers vapour pressure of solvent in proportion to amount of substance dissolved.

Baboon. Large monkey having long dog-like muzzle and large lips.

Backcross. A genetic cross in which a heterozygous organism is crossed with one of its homozygous parents. Thus two backcrosses are possible.

Back E.M.F. Of Cell. An *E.M.F.* set up in a *cell* that opposes the normal *E.M.F.* It occurs due to polarization.

Back E.M.F. Of Electric Motor. An *E.M.F.* set up in the coil of an *electric motor*, opposing the current flowing through the coil, when the *armature* rotates.

Background (phys.). The counting rate of a *counter tube* caused by sources other than the one being measured. It is primarily due to natural *radioactivity* in the soil and *cosmic rays.*

Backing Storage. *Computer stores* with a capacity to store enormous quantities of information, but with an *access time* much greater than the main store. The commonest types are *magnetic* tape decks, fixed magnetic disk stores and exchangeable magnetic disk stores.

Bacteria. Unicellular organisms with a diameter of 1-2 microns. Some bacteria cause disease, for example, tetanus but others are useful, for example, as sources of antibiotics.

Bactericide. A substance that kills *bacteria.*

Bacteriology. The study of *bacteria.*

Bacteriophage. A *virus* that requires a *bacterium* in which to replicate.

Bacterium. A cellular *micro-organism* incapable of photosynthesis they are usually single celled.

Bakelite. Trade name for various synthetic *resins* of which *phenol-formaldehyde resins* are amongst the most widely known.

Baking Powder. A *mixture* that produces *carbon dioxide* gas, CO_2, on wetting or heating, thus causing the formation of bubbles in dough and making it 'rise'. It usually contains $NaHCO_3$, and *tartaric acid* or *cream of tartar.*

Balance. Apparatus used for weighing; common balance have two pans hanging from beam supported at the centre, but there are several variations depending upon purpose and accuracy desired; super-sensitive balance (micro-balance) can weigh upto the nearest 0.0000001 gram.

Balanced Diet. The exact nutritional components required for health generally used in reference to human beings and domesticated animals.

Balata. A natural *rubber*-like material quite similar to *gutta-percha.*

Ballistic Galvanometer. An instrument for measuring electric *charge* by detecting a momentary current. Any *galvanometer* may be used ballistically provided that its period of oscillation is quite long compared with the time during which the current flows.

Ballistic Missile. ·A ground-to-ground-missle having a parabolic flight path. A missile which is propelled and guided only during the initial phase of its flight.

Ballistic Pendulum. A device used for measuring the *ielocity* of a *projectile*. It consists of a large *mass* freely suspended from a horizontal beam and a means of measuring the displacement of the mass when it is struck by the projectile.

Ballistics. It is the study of the flight path of *projectiles.*

Balmer series. A series of sharp distinct lines in the visible *spectrum* of hydrogen.

Band Spectrum. An *emission* or *absorption spectrum* consisting of a number of bands each having one sharp edge. Each band consists of a large number of closely spaced lines. Band spectra arise from *molecules.*

Bandwidth. The range of *frequencies* within which the performance of a *circuit, receiver*, or *amplifier* does not differ from its maximum value by a specified amount. The bandwidth of a *radio* emission is the width of the *frequency band* that carries a specified proportion (usually 99%) of the total *power* radiated.

Bar. A unit of *pressure* in the *c.g.s., system.* It is equal to a pressure of 10^6 *dynes* per sq. cm., or a pressure of 0.986 923 *atmosphere* (approx. 750 mm Hg.) I bar=10^5 *Pascals* commonly used millibar is equal to 100 Pa.

Barbitone ($C_8H_{12}N_2O_3$). White crystalline solid obtained from barbituric acid, melting point 191°C; its sodium salt is used as hypnotic.

Barbituric acid ($C_4H_4N_2O_3$). White crystalline powder obtained by action of malonic acid and urea, melting point 248°C. It is source of important sedatives and also used in plastics.

Barbule. Minute filaments in a row on each side of baro, barbules of adjacent barbs generally hook together and link barbs into a firm vane.

Barff Process. It is a process used for prevention of rusting of iron by the action of *steam* upon the surface of the red-hot metal. It results in a surface coating of black oxide of iron, Fe_3O_4.

Barium. Ba. Element. R.a.m. 137.3. At. No. 56. A silvery-white soft metal. It tarnishes readily in air. R.d. 3.5, m.p. 710°C, b.p. 1640°C. It occurs as *barytes*, $BaSO_4$, and as *barium carbonate*, $BaCO_3$. Its compounds are used in the manufacture of *paints, glass* and fireworks.

Barium Carbonate. $BaCO_3$. A heavy white poisonous. Its powder insoluble in water used in rat poisons, certain optical glasses, and various industries.

Barium. Hydroxide Caustic Baryta. $Ba(OH)_2 \cdot 8H_2O$. A white poisonous solid, m.p. 76°C, used for recovering sugar from waste molasses, for refining *vegetable oils* and in *glass* manufacture.

Barium Oxide Baryta. BaO. A white powder, m.p. 1923°C. It is used as a dehydrating agent and in the manufacture of *glass*.

Barium Peroxide. BaO_2. A white powder, m.p. 450°C, used as a bleaching agent.

Barium Sulphate Blanc fixe. $BaSO_4$. A white crystalline powder, m.p. 1580°C, used as a *pigment* and, as the basis of 'barium meal' in X-ray diagnosis.

Barium Titanate Ba TiO_3. A crystalline substance having good *ferroelectric* and *piezoelectric* properties, used in *transducers*.

Barkhausen Effect. The effect observed when a *ferromagnetic substance* is magnetized by a slowly increasing *magnetic field*; the magnetization does not take place continuously, but in a series of small steps. The effect is due to orientation of *magnetic domains* present in the substance.

Barn. A unit of area for measuring the *cross-section* of *nuclei.* 1 barn equals 10^{-24} sq. cm.

Barnet Effect. Production of small magnetisation in unmagnetised iron rod when it is rotated at very high speed around its axis.

Barometer. An instrument used for measuring atmospheric pressure.

Baroscope. An instrument which indicates changes in density of air.

Barytes. Mineral form of barium sulphate.

Base. Any substance, ionic or molecular, capable of accepting a proton (hydrogen ion).

Bass. Sound waves of frequencies at lower end of audio range (below 250 hertz).

Basic Slag. An impure *mixture* of tetracalcium phosphate $Ca_4P_2O_9$, *calcium silicate* $CaSiO_3$, *lime*, CaO, and *iron (III) oxide*, Fe_2O_3. A by-product of *steel* manufacture, it is a valuable *fertilizer.*

Bath Salts. Na_2CO_3. $NaHCO_3$. $2H_2O$, or some other *soluble* sodium *salt* used to soften the *water.*

Bastnasite. A rare yellow-brown *mineral.* It consists of a *carbonate* of fluorine and several lanthanides, *i.e.* $LaFCO_3$. It is used as a source of lanthanides, including praseodymium.

Bathymetry. Measurement of depth, especially of the sea.

Battery. A number of *primary* or *secondary cells* arranged in series or in parallel. In series, they give a multiple of the E.M.F. of the cell; in parallel, they give the same E.M.F. as the cell, but have a greater capacity, *i.e.*, a given current can be supplied for a longer period. The common dry batteries' usually consist of *Leclanche cells.*

Baume Scale. A scale of relative *density* (specific gravity) of *liquids* Named after A. Baume (1728-1804).

Degree Baume = 144.3 (r.d.— 1)/r.d.

Bauxite. Natural *hydrated aluminium oxide*, Al_2O_3. xH_2O. The most important *ore* of aluminium.

Bauxite Cement. Ciment fondu. A rapid-hardening *cement* consisting mainly of calcium aluminate; made from *bauxite* and *lime* in an electric furnace.

Beam (Phys.) *Radiation* travelling in a particular direction.

Beam Hole. A hole made in the shields and usually through the *reflector,* of a *nuclear reactor*. It permit, the escape of a beam of radiation, particularly *neutrons*, for experimental purposes.

Beam Riding. A method of *rocket* guidance in which the missle steers itself along the axis of a *beam* of *radiation.*

Beam Transmission. *Radio* transmission in which the *electromagnetic waves* are sent in a particular direction in a *beam* and not being radiated in all directions.

Bearing (math). The direction of a point *B* from a fixed point *A* in terms of the angle the line *AB* makes with the line running due **North** through *A*, considered in a clockwise direction.

Beat Frequency. The difference *frequency* arising out from the interaction between *radio frequency* signals of different *wavelengths.*

Beats (phys.) A periodic increase and decrease in loudness heard when two notes of nearly the same *frequency* are sounded simultaneously.

Beaufort Scale. A numerical scale for the estimation of wind force, based on its effect on common objects.

Beckmann Thermometer. A sensitive *thermometer* for measuring small difference or changes in *temperature.* The scale covers 6-7 degrees and is graduated to 0.01 degree.

Becquerel. The derived *SI unit of activity* (radioactive). The activity of a *radionuclide* that decays at an average rate of one spontaneous nuclear transition per second. 1curie=3.7×10^{10} Bq.

Beeswax ($C_{30}H_{61}O$. $C_{15}H_{31}$). Greyish-brown wax having yellowish tinge consisting of mixture of compounds secreted by bees for constructing their honey-combs; used in cosmetics, pharmaceutics, polishes etc.

Benzaldehyde (C_6H_5CHO). Oil of bitter almonds. It is a colourless liquid used in dye industry and perfumery.

Benzene (C_6H_6). Aromatic hydrocarbon discovered by Faraday in 1825 can be prepared by destructive distillation of coal-tar; It is used as solvent for fats, resins etc, and in drycleaning of clothes.

Benzine. Petroleum fraction. It is a mixture of alkanes from petroleum boiling between 35 and 80°C. It is used as a solvent.

Benzoic acid. While solid soluble in alcohol. M.P. 395°C.

Benzoline. Refers to impure benzene.

Benzonitril (C_6H_5CN). It is a colourless poisonous liquid, B.P. 190.7°C.

Benzopyrene ($C_{20}H_{12}$). Polycylic hydrocarbon. Found in coal tar in small quantities.

Beri-Beri. Deficiency disease caused due to deficiency of vitamin B_1.

Beta Decay. Radioactive decay involving emission of an electron.

Beta Factor. In plasma physics the ratio of plasma kinetic pressure to magnetic pressure.

Beta-Iron. Allotropic form iron, similar to alpha-iron, it is non-magnetic, stable between 768° and 910°C.

Beta Rays. Stream of beta particles. They have large penetrating powers.

Betatron. Apparatus used for obtaining particles of high energy by keeping beam of accelerated electrons moving in a stable orbit.

BHC ($C_6H_6C_{16}$). Benzene hexachloride. Greyish amorphous solid having musty smell; used as contact and stomach poison against a wide spectrum of insect pests.

Bi-concave. Lens which is concave on both sides.

Bi-convex. Lens with both side convex.

Bile. A green alkaline fluid produced in the liver of mammals. Bile is stored in the gall bladder and is transported via the bile duct to the duodenum where it causes emulsification of lat prior to the digestion.

Binary Compound. Chemical compound of two elements only (ending with-ide)—calcium chloride etc.

Binary Notation. A system of numbers having only two different digits, usually 0 and 1. The notation is used in computers.

Binary Stars. System of two stars gravitationally attracted to each other and revolve round common centre of gravity.

Binocular. Optical device used for simultaneous vision with both eyes.

Bioassay. Quantitative estimation of biologically active substance.

Biochrome. Any naturally occurring animal or plant pigment.

Biocide. General poison or toxicant.

Biodegradation. It is the chemical degradation and causes environmental pollution. Biodegradable substances such as sewage, cause no pollution if correctly treated. Some substances which are non-biodegradable cause environmental pollution *e.g.*, some insecticides and some heary metals.

Biogen. Hypothetical protein molecule which is unstable but is having essential properties of living thing.

Biogenesis. Refers to theory that living things are derived only from living beings.

Biogenetic Law. Doctrine that during development of higher animals there appear stages similar to those present in adults of lower type ancestors.

Biogeography. Means geographical distribution of plants and animals.

Biological Control. Control of pests by means of parasites, predators and disease producing organisms.

Biological Warfare. Intentional use of disease causing organisms for destroying crops and animals (including human beings).

Bioluminescence. Refers to the production of light by living creatures like glow-worms and fireflies; light is emitted by oxidizing luciferm in presence of enzyme luciferase; also found in many marine organisms.

Biophysics. The branch of science which deals with the application of physics to the study of biology.

Bioscope. Cinematographic apparatus used for projecting movie films.

Biosphere. World of living organisms together with layers of soil, water and air.

Biosynthesis. Synthesis of chemical compounds with the help of living organisms.

Biotin($C_{10}H_{16}O_3N_2S$). Heat stable coenzyme, one of vitamin B-complex group; It is also known as vitamin H.

Biotype. Individual species having same genetic characteristics.

Biphenyl. Diphenyl ($C_{12}H_{10}$). A colourless insoluble powder used in manufacture of organic dyes.

Biphenyl-4,4'-Diamine—Benzidine $NH_2C_6H_4.C_6H_4NH_2$. An aromatic base used in dye industry.

Biplane. Aeroplane or glider having two sets of wings, one above the other.

Biprism. Two acute-angled prisms placed base to base; It is used for obtaining interference fringes.

Birkland and Eyde process. A process for fixation of atmospheric nitrogen.

Bisection. To divide in two parts.

Bisector. A straight line which divides other line or angle in two equal parts.

Bismuth (Bi). Silvery-white metal having reddish tinge, it is brittle and poor conductor of heat and electricity, boiling point 156°C.

Bittern. Oily liquid that is left after crystallisation of salt from sea-water.

Bitumen. Refers to various inflammable mineral substances like asphalt, naphtha, petroleum etc.

Bituminous. Substance impregnated with bitumen *e.g.*, coal that contians 78 to 80% carbon, is hard like stone and burns with smoky flame.

Biuret. $NH_2CONH.CONH_2.H_2O$. A crystalline insoluble product obtained from urea. It is also known as carbamoyl urea.

Biuret Reaction. An alkaline solution of Biuret reacts with 1% $CuSO_4$ solution in the presence of peptide bonds to give a purple colour. This is an addition reaction and used as a test for protein and urea.

Bivalent. Having valency of two.

Black Ash. Impure sodium carbonate.

Black Body. One that will absorbs all incident radiation and reflects none.

Black Body Radiation. Emission of radiant energy that would take place from black body at fixed temperature.

Black Light. Invisible light, such as ultraviolet rays that fall on fluorescent materials which in turn emit visible light.

Black Damp. Refers to carbon dioxide found in coal mines.

Black Hole. A region (hypothetical) in space from which no matter or radiation can escape because of its intense gravitational field. These regions are formed when stars collapse.

Black Lead. Graphite. A natural form of carbon.

Blanc Fixe. Refers to artificial $BaSO_4$. It is commonly used as an extender in paint industry.

Bland. Chemical that causes no discomfort *e.g.*, olive oil.

Blanket. A layer of fertile material which surrounds the core of a nuclear reactor and acts as a reflector.

Blast Furnace. Furnace used for smelting of iron.

Blasting Gelatin. A mixture of gun cotton and nitroglycerine. It is a strong explosive.

Bleach. To whiten; remove the colour chemically from coloured materials.

Bleaching Agent. Any oxidising or reducing substance that can be used for bleaching —chlorine, hydrogen peroxide, sulphur dioxide etc.

Bleaching Powder. Calcium oxychloride ($CaOCl_2$) is white powder which is prepared by action of chlorine on calcium hydroxide [$Ca(OH)_2$]; also called chlorinated lime.

Bleeder. Any individual afflicted with haemophilia.

Blende. Sulphide ore occurs in nature.

Blind Spot. Area of retina in eye where optic nerve leaves the eyeball; it is sensitive to light.

Blotting Paper. Porous paper having innumerable cappillaries for sucking up ink or any other liquid by virtue of surface tension.

Blown oil. A thickened *oil* which is made by blowing air through a natural vegetable or animal oil.

Blowpipe. A device for producing a jet of *flame* by forcing a flammable *gas* mixed with air or oxygen through a nozzle at high pressure.

Blue Stone. See *blue vitriol.*

Blue Vitriol. Bluestone. Crystalline *copper (II) sulphate* pentahydrate, $CuSO_4.5H_2O$. Used for copper plating and in *solution* for spraying plants.

Board of Trade Unit. B.O.T. unit. A former name for the *kilowatt-hour.* The energy obtained when a *power* of kilowatt is maintained for 1 hour.

Boart. See *bort.*

BOD. Biochemical (or biological) oxygen demand. A measure of the content of organic matter in water and wastes. It is the amount of oxygen (mg of O_2 per cubic decimetre of water) when a sample containing a known mass of oxygen in solution is kept at 20°C. for five days. The oxygen is consumed by microorganisms that feed on the organic matter in the sample.

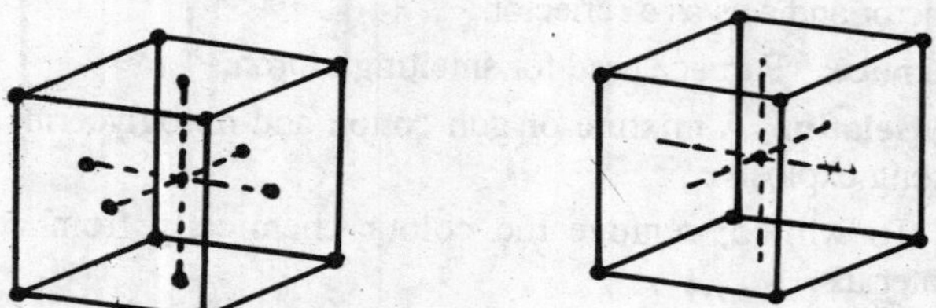

BODY-CENTRED CUBIC CRYSTAL

FACE-CENTRED CUBIC CRYSTAL

Fig. 1.

Body-Centred. Means for a crystal in which there is a *lattice* point at the centre of the body of the crystal as well as at the corners. It is said to be 'face-centred' when a lattice point is present at the centre of each face. See Figure 1.

Bog Iron Ore. An impure form of *hydrated iron (II) oxide*, $Fe_2O_3.xH_2O$, found in bogs and marshes.

Bohr Atom. Atomic model having structure postulated in Bohr theory.

Bohr Orbit. One of the electron paths around nucleus in Bohr's model for hydrogen theory.

Bohr Radius. Refers to the radius of ground state orbit of hydrogen atom in Bohr theory.

Bohr Sommerfeld Theory. Modification of Bohr theory in which circular as well as elliptical orbits are permitted.

Bohr Theory. Theory of atom put forwarded by Niels Bohr to explain line spectrum observed for hydrogen.

Boiling point. Refers to the temperature at which liquid boils freely under standard atmospheric pressure; boiling point of water is 100°C. Temperature at which vapour pressure of liquid becomes equal to atmospheric pressure.

Bomb calorimeter. Calorimeter used for finding out heat of combustion.

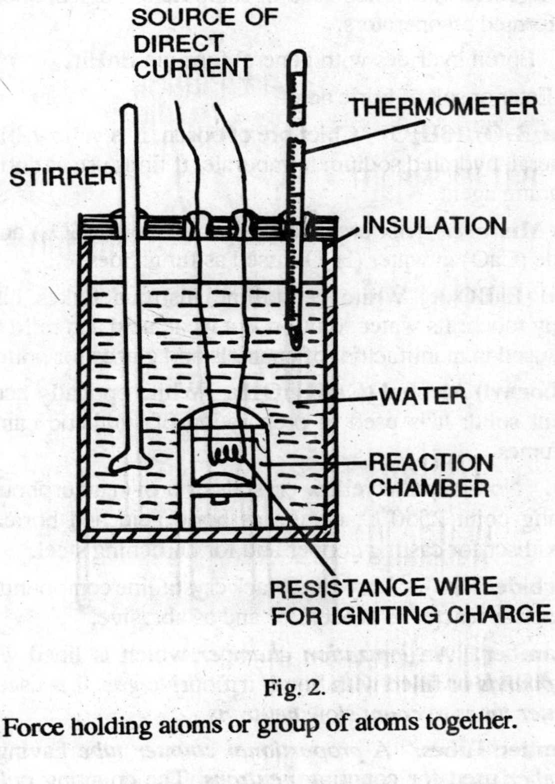

Fig. 2.

Bond. Force holding atoms or group of atoms together.

Bond Energy. Energy required for seperating a particular bond in an specified substance.

Bond Length. Distance between nuclei of two atoms joined by chemical bond.

Bond Polarity. Difference in ability of two atoms in a chemical bond to attract electrons.

Bone Ash. *Ash* obtained by heating bones in air. It consists mainly of *Calcium phosphate* (*V*), $Ca_3(PO_4)_2$.

Bone Black. See *animal charcoal.*

Bone Char. See *animal charcoal.*

Bone Oil. Dippel's oil. A product obtained by the destructive distillation of bones. It is a dark oily evil-smelling liquid used as a source of *pyridine.*

Booleun Algebra. A branch used in computers. Logical operations are performed by operators.

Boranes. Boron hydrides with general formula $BnHn_{+2}$

Borate. Ester or salt of boric acid.

Borax ($Na_2B_4O_7.10H_2O$) Chief ore of boron, it is yellow, blue or grey mineral, hydrated sodium tetraborate. It finds use as antiseptic and cleaning agent.

Bordeaux Mixture. Mixture of copper sulphate ($CuSO_4$) and calcium oxide (CaO) in water (H_1O); used as fungicides.

Boric Acid (H_3BO_3). White crystalline lustrous flakes having soft soapy touch; Its water solution is a weak acid and mild antiseptic. It is used in manufacture of enamels and glazes for pottery.

Borneol (bornyl) Alcohal ($C_{10}H_{17}OH$). White optically active translucent solid. It is used in preparation of synthetic camphor and perfumes.

Boron (B). Non-mettalic yellow crystals or brown amorphous powder, boiling point 2550°C; occurs as boric acid and borax; used as deoxidiser for casting copper and for hardening steel.

Boron Carbide (B_4C). Very hard black crystalline compound; It is used in nuclear reactor as moderator and as abrasive.

Boron Chamber. An *ionization chamber* which is lined with boron *compounds* or filled with boron triflouride *gas*. It is used in boron *counter tubes* to *count slow neutrons.*

Boron Counter Tubes. A *proportional counter tube* having a *boron chamber* used for counting ***neutrons*. The** counting pulse results from particles emitted when *neutrons* react with the boron-10 *isotope.*

Boron Nitride (BN). Slippery white solid which is obtained by action of nitrogen on boron. It exists in cubic and hexagonal forms: It is used as lubricant, electric insulator for high voltage and thermal insulation of furnaces.

Boron Trichloride (BCl_3). Colourless mobile liquid, boiling point 18°C. It fumes in moist air because it gets readily hydrolysed by water; used in extraction of boron.

Boroh Triffouride (BF_3). Colourless gas in dry atmosphere that gives white dense fumes in moist air. It is used for esterification of alcohols.

Borosilicates. A group of silicates in which the SiO_4 units are linked with BO_3 units in a variety of structures. The borosilicates glasses, such as *Pyrex*, have a smaller plastic range and are resistant to chemical attack.

Bort. Boart. Impure or discoloured diamond; useless as a gem. However, it is as hard as pure diamond and is used for drills, cutting tools, etc.

Bosch Process. An industrial process for the manufacture of hydrogen. *Water gas*, a *mixture* of *carbon monoxide* and hydrogen, is mixed with *steam* and passed over a heated *catalyst*. The steam reacts chemically with the carbon monoxide to give *carbon dioxide* CO_2, and hydrogen. The CO_2 is then removed by dissolving it in water under pressure. Named after C. Bosch (1874—1940).

Bose—Einstein Statistics. The branch of *statistical mechanics* used with systems of identical particles having the property that the *wave function* remains unchanged if any two particles are interchanged. See *boson*. Named after S.N. Bose (1894-1974) and Albert Einstein (1879-1955).

Bosons. *Elementary particles*, such as *photons* and *mesons*, that conform to *Bose-Einstein statistics*; their numbers are not conserved in particle interactions. Bosons have *integral spin* (0,1,2). Compare *fermions.*

Boyle's Law. For an ideal gas at a constant temperature, volume of gas is inversely proportional to its pressure.

Brass. Alloy of copper and zinc.

Brewster's Law. The tangent of the angle of *polarization* is numerically equal to the *refractive index* of the reflecting medium when the polarization is maximum.

Brimstone. Sulphur fused into blocks.

Brinell Test. A test for the hardness of *metals*. A ball of chrome *steel*, or other hard material, of standard size, is pressed by a heavy load into the surface of the metal, and the diameter of the depression is measured. The Brinell Number is the ratio of the load in *kilograms* to the *area* of the depression in square millimetres.

Bristle-tail. Small primitive wingless insect; It has spindle-shaped body covered with scales, antennae long and filiform, cerci long and multisegmented.

Britannia Metal. An *alloy* of variable composition, containing 80%-90% tin, with some antimony and copper, and sometimes also zinc and lead.

British Thermal Unit. The quantity of *heat* required to raise the *temperature* of 1 lb of *water* through 1° Fahrenheit; it is equal to as 1055.06 *joules*.

Brittle. Solid material that breaks into small pieces on applying force or pressure *e.g.*, glass.

Bromate. Salt of bromic acid.

Bromic Acid ($HBrO_3$). Compound of hydrogen, bromine and oxygen.

Bromide. Binary compound having bromine. Salt of hydrobromic acid, potassium bromide (KBr), silver bromide (AgBr) etc.

Bromide Paper. Photography paper having one side sensitive surface coated with silver bromide; used for printing photographs from negatives.

Bromine (Br). Non-metallic dark red fuming liquid. It has powerful chlorine- like smell, poisonous brown vapour, boiling point 59°C, fairly soluble in water and very soluble in alcohol, ether and carbon disulphide; used as disinfectant and in manufacture of dye-stuffs, bromides, bromites, bromates etc., as also oxidising agent in organic synthesis.

Bronze. Alloy of copper and tin.

Brownian Movement. Random movement of small particles in a fluid; caused by continuous bombardment of particles by molecules of that medium.

Brucite. A mineral consisting of *magnesium hydroxide*.

Brush Discharge. An electric discharge from sharp points on a *conductor.* The surface density (*i.e.*, quantity of electric charge per unit area) is greatest at sharp points; the high charge at such points causes a displacement of the charge on the air particles near the points, and hence an attraction to the points. On reaching the points, the particles acquire some of the charge on the points and are repelled. This causes a stream of charged air particles to leave the vicinity of the points.

Bubble. A thin liquid film that forms a ball around air or glass.

Bubble Chamber. An instrument for making the tracks of ionizing particles visible as a row of bubbles in a *liquid.* The liquid is heated to slightly above its *boiling point* and maintained under *pressure* to prevent boiling.

Buchner Funnel. Funnel made of procelain having flat circular base perforated with small holes; used for filtering by suction.

Bunsen Burner. Simple gas-burner with an adjustable air-valve. Proper mixing of air with gas before ignition which results in smokeless flame of low luminosity but great heating power.

Buoyancy. Upward thrust experienced by a body when it is immersed in a liquid.

Burette. Graduated long narrow glass tube fitted with glass tap at one end and open at other end. It is used for volumetric analysis— various types like single bore, double bore, overflow cup, pressure filling etc.

BURETTE

Fig. 3.

Butadiene (C_4H_6). Hydrocarbon gas which is used in manufacture of synthetic rubber.

Butanal [$CH_3(CH_2)_2CHO$]. Colourless inflammable liquid. It is used in plastic and rubber industries.

Butane (C_4H_{10}). Hydrocarbon of methane series; colourless gas obtained by electrolysis of sodium acetate and sodium propionate. It is used as portable supply of fuel.

Butanone($C_2H_5COCH_3$). Inflammable liquid, boiling point 80^oC. It is used as solvent in manufacture of plastics.

Butyl (C_4H_9). Hydrocarbon radical derived from butane.

Butyl-Rubber. Synthetic rubber.

Butylene (C_4H_6). Olefine hydrocarbon in three isomers.

Butyric Acid ($C_3H_7.COOH$). Colourless, volatile, fatty liquid smelling like rancid butter, miscible with water, ethanol and ether.

Bypass Capacitor. Bypass condenser. A *capacitor* that provides a path of low *impedance* over a certain range of *frequencies.*

By-Product. A substance obtained incidentally during the manufacture of some other substance. It may be as important as the manufactured substance itself.

Boyte. A single unit of information handled by a *computer*; usually 8 *bits*.

C

Cache Memory. Extremely last semiconductor memory that is used in high performance computer systems in association with main source.

Cacodyl $(CH_3)_2$ As. Dimethyl arisino group.

Cadmium (Cd). Soft silvery-white metal. It occurs in very small amounts (0.2 to 0.5%) in zinc ores. Its boiling point is $765^{\circ}C$. It is good absorber of neutrons; used in manufacture of control rods for nuclear reactors.

Cadmium Carbonate ($CdCO_3$). White amorphous powder can be obtained by adding any soluble carbonate to solution of cadmium salt; It is used as laboratory reagent to detect presence of sulphide ion.

Cadmium Oxide (CdO). Dark brown powder which is obtained by heating cadmimum or igniting its carbonate or nitrate, insoluble in water but soluble in acid forming cadmium salts.

Cadmium Sulphide (CdS). Bright yellow amorphous powder which is obtained by passing hydrogen sulphide into solution of cadmium salts; insoluble in water and dilute acids but dissolves in concentrated acids.

Caecum. The part of the mammalian gut at the entry to the large intestine. In man it has no important function but in herbivores, it is an important site of cellulose digestion.

Caesium (Cs). Soft silvery-white metal, boiling point $690^{\circ}C$, it is very reactive; used in manufacture of radio-tubes and photoelectric cells.

Caesium Chloride (CsCl). Typical alkalihalide obtained from caesium carbonate (Cs_2CO_3) and hydrochloric acid (HCl).

Caesium Clock. Used in S.I. unit definition of a *second*. It constitutes an extremely accurate clock. It is based on energy difference between two states of Cs nucleus in a magnetic field. This energy difference corresponds to a frequency of 9192631770 *hertz*.

Caffeine ($C_8H_{10}O_2N_4$). Silky white crystalline purine that occurs in tea leaves, coffee beans etc. It has powerful action on heart and is used in various medicines.

Calamine. Zinc mineral used for skin preparations, its lotion or powder is also used for sunburn.

Calciferol. Vit. $D_2C_{28}H_{43}OH$—unsaturated alcohol. Controls the deposition of calcium compounds in body. Its deficiency causes rickets.

Calcination. The process of heating an ore to a high temperature to expel water from hydrates or hydroxides or carbon dioxide from carbonates.

Calcite. Natural crystalline form of calcium carbonate, occurs in marble and limestone.

Calcium (Ca). Hard, silvery-lustrous metal. It is malleable and ductile, found in chalk, lime, gypsm etc.

Calcium Carbide (CaC_2). Grey crystalline solid, prepared by heating lime and powdered coke in electric furnace (3000°C); It is used as fertilizer.

Calcium Cyanamide ($CaCN_2$). Grey crystalline powder obtained by heating calcium carbide in atmosphere of nitrogen at 1000°C; It is used as fertilizer (urea), it is converted by soil moisture into ammonia.

Calcum Cyclamate. $(C_6H_{11}NHSO_3)_2Ca.2H_2O$ white powder, soluble in water. Previously used as a sweetening agent in soft drinks.

Calcium Carbide. CaC_2 greyish solid, reacts with water to give acetylene.

Calcium Hydroxide [$Ca(OH)_2$]. Slaked lime; white crystalline powder can be obtained by action of water on calcium oxide; It is used in cement, mortars, plasters etc.

Calcium Oxide (CaO). Quick lime; white amorphous powder can be obtained by heating limestone in vertical rotary kiln (1000°C); It is used for making cement and mortar.

Calcium Phosphate [$Ca_3(Po_4)_2$]. White precipitate prepared by addition of soluble phosphate to alkaline solution of calcium salt; It is used in manufacture of calcium super phosphate phosphoric acid and phosphorous.

Calcium Silicates. Used in glass and cement industry. Various Calcium silicate phases are formed in the process.

Calcium Sulphide CaS. Colourless substance having smell of bad eggs.

It is used to prepare luminous paints and cosmetics.

Calculus. A branch of mathematics and is a powerful method of solving mathematical problems.

Calibration. The *graduation* of an instrument to enable measurements in definite *units* to be made with it.

Caliche. Impure *sodium nitrate* $NaNO_3$. found in Chile.

Californium. **Cf.** *Transuranic element*. At. No. 98. Its most stable *isotope*, californium-251, has a *half-life* of 800 years.

Callipers. It is an instrument for measuring the distance between two points on a curved surface; *e.g.*, for measuring the internal and external diameters of tubes.

Callose. A carbohydrate deposit found seasonally or permanently on sieve plates. It causes them to stop functioning.

Calomel Electrode. A *half cell* consisting of a mercury electrode covered with *calomel* (mercury (I) chloride) and a solution of mercury in *potassium chloride*. It is used as a standard electrode.

Calorescence. Absorption of *light* radiations by a surface, their conversion into *heat* and the consequent emission of heat *radiation.*

Calorie. Unit of quantity of *heat.* It is the quantity of heat required to increase the *temperature* of 1 g of *water* through 1°C. The International Calorie is defined as 4.1868 joules. The joule is the *SI unit* of *heat* and has largely replaced the calorie.

Calorie, Large. Kilogram-calorie. 1000 *calories.* Written Calorie or Kcalorie, it is still used for quoting energy values of foods but is becoming outdated.

Calorific Value. The quantity of *heat* produced by a given *mass* of a fuel on complete *combustion*. It is expressed in *joules* per *kilogram* (*SI units*) or frequently megajoules per kilogram. Calorific values are determined by using a *bomb calorimeter.*

Calorimeter. An instrument used for determining quantities of *heat* evolved, absorbed, or transferred.

Calorimetry. Dealing with the measurement of heat.

Calx. Powdery metal oxide formed on roasting an ore or a mineral.

Calyx. Refers to outer floral circle of green coloured sepals.

Camel. A desert animal.

Camera. A device used for obtaining photographs.

Camphor ($C_{10}H_{16}O$). White crystalline solid. It has characteristic smell. It is used as carminative and stimulant and also in manufacture of celluloid.

Canada Balsam. Yellowish fluid having refractive index similar to that of glass; It is used for mounting microscopic slides and as an adhesive for optical instruments.

Canal Rays. Positively charged *ions* produced during an *electric discharge* in gases. They are accelerated towards the *cathode* by the applied *potential difference* and allowed to pass through canals bored in the cathode.

Candela. The *SI unit of luminous intensity*. The luminous intensity, in the perpendicular direction, of a surfae of 1/600,000 square *metre* of a *black body* at the *temperature* of freezing platinum under a *pressure* of 101 325 Nm^{-2}. Symbol cd.

Candlepower. It is the *luminous intensity* of the source of light in that direction expressed in terms of the *candela.*

Candle Wax. The wax used to make candles, usually either *paraffin wax* or *stearine.*

Cane Sugar. Sucrose, saccharose. $C_{12}H_{22}O_{11}$. A *disaccharide* which is obtained from the sugar-cane. It is chemically indentical with *beet sugar.*

Cannibal. Refers to one who eats flesh of his own species.

Cantou's Phosphorous. CaS. Impure and has the property of phosphorescence on exposure to light used in preparation of luminous paints.

Caoutchone. Used for Raw rubber.

Capacitance. Electrical capacity. *C*. The property of a system of electrical *conductors* and *insulators* which enables it to store *electric charge* when a *potential difference* exists between the conductors. The *SI unit* of capacitance is the *farad.*

Capacitor. Electrical condenser. A system of electrical *conductors* and *insulators*, the principal characteristic of which is its *capacitance.* The simplest form consists of two parallel *metal* plates separated by a layer of air or some other insulating material.

Capacitor Microphone. Condenser microphone. A *microphone* consisting essentially of an electrical *capacitor*, one plate of which is fixed and the other plate forms the diaphragm upon which the *sound* waves fall.

Capillary Action. Capillarity. The phenomena observed in *liquids* due to inter molecular attraction at the liquid boundry; *e.g.*, the rise or depression of liquids in narrow tubes, the formation of films, drops, bubbles, etc.

Capillary Tube. A tube of very small internal diameter.

Capture. Refers to a process by which an atomic or nuclear system acquires an additional particle.

Caramel. Brown substance formed by heating sugar. It has a complex composition.

Carat. Measure of weight of diamonds (=200 miligrams). Fineness of gold expressed as 24 parts of alloy *i.e.*, 24 carat gold is pure gold.

Carbamoyl. Refers to NH_2CO — radical.

Carbanion. R_2C^- a negative ion containing carbon atom. It occurs as intermediate in many organic reactions.

Carbaryl. **($C_{12}H_{11}NO_3$)** white solid, insoluble in water. It is used as a contact and systemic insecticide.

Carbendazim. Methyl benzimidazole (MBC). It is used as a systemic fungicide.

Carbene. Organic group R_2C;, H_2C: is the simplest carbene.

Carbocyclic Compounds. Closed ring organic compounds.

Carbohydrases. Enzymes. These hydrolyse carbohydrates. Examples are amylase, maltase etc.

Carbohydrates. Organic compounds with general formula $C_x(H_2O)_y$. They are an essential part of metabolism of all living organism.

Carbon. C Most widely distributed non-metal.

Carbonado. Impure variety of diamond. Being black useless as a gem. Being hard it is used for drills.

Carbonate. A *salt* of *carbonic acid*, H_2CO_3.

Carbonation. Treatment with *carbon dioxide*, usually done for the formation of *carbonates.*

Carbon Black. A finely divided form of *carbon*, produced by *pyrolysis* or by incomplete *combustion* from carbon-rich materials, such as *mineral oils*, *ethyne*, or *natural gas*, used mainly as a reinforcing *pigment* in rubber, and also as a black pigment in inks, plastics, etc.

Carbon Cycle (bio). The circulation of carbon between living *organisms* and the *atmosphere*. Carbon dioxide is built into complex carbon *compounds* by plants during *photosynthesis*; animals obtain their carbon *atoms* by feeding on plants or other animals; during *respiration*, and by decay after death, some of this carbon is returned to the atmosphere in the form of carbon dioxide.

Carbon Cycle (phys). A cycle of six consecutive *nuclear reactions* resulting in the formation of a *helium* nucleus from four *protons.*

The carbon nuclei with which the cycle starts are reformed at the end and therefore act as a *catalyst*. The energy liberated by the carbon cycle is thought to be the main source of *energy* in a large class of *stars*.

Carbon Dioxide. **Carbonic acid gas. Co$_2$.** A colourless, odourless and tasteless gas. It occurs in the *atmosphere* as a result of the *oxidation* of carbon and carbon *compounds*. Atmospheric carbon dioxide is the source of carbon for plants. It solidifies at —78.5°C. at atmospheric pressure, and is used as a *refrigerant* in solid form as *dry ice,* for the preservation of frozen foods, etc., it is used in *fire extinguishers.*

Carbon Disulphide. **Carbon bisulphide. CS$_2$.** A colourless flammable liquid, b.p. 46°C. It has a high *refractive index*. It is made by heating sulphur with carbon or with *methane* at high temperatures. It is used as a *solvent* in various industrial processes and in manufacture of viscose *rayon*, used as a *pesticide.*

Carbon Fibre. A material consisting of black silky threads of pure carbon that can be made stronger and stiffer than any other material of the same *weight*. They are used to reinforce a matrix of *resin, ceramic*, or *metal* with up to 600,000 fibres per square centimetre of cross-section and in this form make a valuable constructional material where strength is required at high *temperature*, such as in components for jet engines and *rockets.*

Carbonic Acid. **H_2CO_3.** A very weak *acid* formed in small amounts when *carbon dioxide* dissolves in *water*. It is never obtained pure as it breaks up almost completely into carbon dioxide and water when obtained in a chemical reaction. It gives rise to two series of *salts*, the *carbonates* and bicarbonates.

Carbonic Anhydrase. Enzyme which catalyses reaction between carbon dioxide and water.

Carbonise. To reduce to carbon. To cover with carbon.

Carbon Monoxide (CO). Colourless, poisonous and tasteless gas. It is produced when carbon is burnt with insufficient amount of oxygen; It is used as fuel.

Carbon Tetrachloride (CCl_4). Heavy, colourless liquid, has sweetish smell, boiling point 77°C, insoluble in water but soluble in ethanol and ether; used in fire extinguisher and solvent for fats, oils, resins etc.

Carborundum. ***Silicon carbide*. SiC.** A dark crystalline *solid*, nearly as hard as *diamond*, used as an *abrasive* and as a refractory material. Obtained by heating *silica*, SiO_2, with carbon in an electric furnace.

Carboxyl Group. The *univalent* group, —COOH, characteristic of the organic *carboxylic acids.*

Carboxylic Acids. *Organic acids* containing one or more *carboxyl groups* in the molecule; *e.g., ethanoic acid,* CH_3COOH. In systematic nomenclature they end in -oic.

Carburettor. A device used in the *internal-combustion* petrol engine for mixing air with *petrol vapour* prior to explosion.

Carcinogen. A substance (that is) capable of producing cancer (carcinoma).

Cardiogram. Tracing of heart beats as recorded by cardiograph.

Cardiograph. Instrument used for recording movement of heart.

Carnallite. Chloride ore of potassium and magnesium.

Carnosine. $C_9H_{14}N_4O_2$. An *optically active* crystalline *dipeptide,* m.p. 260°C., It is found in muscle *tissue.*

Carnotite. An ore of uranium. It consists of uranium potassium *vanadate* of variable composition.

Carnot's Cycle. An ideal reversible cycle of operations for the working substance of a heat engine. The four steps in the cycle are (*a*) *isothermal* expansion, (*b*) *adiabatic* expansion, (*c*) *isothermal* compression, (*d*) *adiabatic* compression.

Carnot's Principle. The *efficiency* of any reversible heat engine depends only on the *temperature* range through which it works and not upon the properties of any material substance. If the heat is taken up at a *thermodynamic temperature* T_1 and given out at a thermodynamic temperature T_2 (as in *Carnot's cycle*), the efficiency is $(T_1 - T_2)/T_1$.

Carotene. $C_{40}H_{56}$. A *yellow unsaturated hydrocarbon* found in carrots and butter. It is converted into vitamin A in the animal *organism.* Carotene acts as a photosynthetic pigment in plant *cells* that lack *chlorphyll.*

Carotenoids. Group of yellow, orange and red plant pigments located in chloroplasts and in plastids in plant parts where chlorophyll is absent.

Carrier. 1. A substance assisting a *chemical reaction* by combining with part or all of the *molecule* of one of the reacting substances to form a *compound* which is then easily decomposed by the other reacting substance; the carrier is thus left unchanged. **2.** An inactive substance used to transport a *radisotope* in *radioactive tracing.* A radioisotope is said to be 'carrier-free' if it can be used without a carrier. **3.** Sometimes called carrier gas. The gas used to carry the sample through the column in gas *chromatography*.

Carrier. Charge carrier. The particles that carry the *charge* when an electric current flows. In a *metal* they are the *free electrons*, in a *semiconductor* they are electrons or *holes*. In an *electrolyte* they are *ions*, and in a gas they are ions and electrons.

Carrier Wave. A continuous *electromagnetic radiation* of constant *amplitude* and *frequency*, emitted by a *radio* transmitter. By *modulation* of the carrier wave, oscillating *electric currents* caused by *sounds* at the transmitting end are conveyed by it to the receiver.

Carron Oil. A mixture of *vegetable oil* (olive or cotton-seed) with *lime-water*. It is used as an application for burns.

Cassette. Small compact container (cartidge) for magnetic tape (audio) which can be recorded and replayed in cassette tape recorder or photographic tape (video) that can be filmed by using special camera for both motion and sound and pre-recorded cassettee can be viewed on television set through VCP (VTP) or VCR (VTR).

Cassiterite (SnO_2). Stannic oxide; principal ore of tin.

Cast Iron (Pig Iron). Alloy of iron containing 2 to 5% carbon and other impurities; very hard and brittle, generally converted into steel or wrought iron but is also used as such in making pipes, radiators, stoves, toys etc.

Catabolism. Part of metabolism dealing with decomposition of complex substances into simpler ones with release of energy.

Catalase. Enzyme used for decomposing hydrogen peroxide.

Catenation. Property of element to link itself to form big molecules as in carbon, phosphorous and sulphur.

Cathetometer. Telescope fitted on graduated verticular pillar along which it can move; It is used for measuring lengths and displacements at a distance of a few meters.

Cathode. Negative electrode in an electrolytic cell; electrode that pulls electrons from external circuit.

Cathode Ray. Stream of electrons emitted from cathode when an electric discharge takes place in vacuum tube.

Cathode Ray Tube Oscilloscope. Instrument (vacuum tube) in which electric signals can be observed on a fluorcescent screen; It is indicator in radar system.

Cathodograph. Photograph by X-rays.

Catholyte. Refers to electrolyte near cathode during electrolysis.

Cation. Positively charged ion obtained by removal of electrons from atoms or molecules.

Caustic. Burning; corrosive towards organic matter. Parallel rays of light falling on concave spherical mirror that do not form a point image at focus.

Caustic Alkali. Sodium or potassium hydroxide.

Caustic Ammonaia. Ammonia as gas or in solution.

Caustic Lime. Quicklime.

Caustic Potash. Potassiumhydroxide (KOH).

Caustic Soda. Sodiumhydroxide (NaOH).

Cavitation. Formation of vapour filled cavities in a flowing liquid on account of decrease in pressure because of high velocity—gas bubbles in liquid.

Cavity Radiator. A hot enclosure having small opening that allows some radiation to escape or enter.

Celestial Equator. The circle in which the *plane* of the Earth's *equator* meet the *celestial sphere.*

Celestial Mechanics. The branch of *astronomy* dealing with the motions of celestial bodies or systems under the influence of *gravitational fields.*

Celestial Sphere. The imaginary *sphere* to the inner surface of which the heavenly bodies appear to be attached;

Celestine. Natural crystalline strontium sulphate, $SrSO_4$.

Cellophane. Transparent sheet cellulose, prepared by extracting cellulose xanthate solution into an acid bath, used as wrapping material.

Celluloid. Thermoplastic material obtained from cellulose nitrate and camphor.

Cellulose. A *polysaccharide* occurs widely in nature in fibrous form as the structural tissue in the cell walls of plants. Its *macromolecules* consist of long chains of *gluclose* units. It can be obtained from wood pulp, cotton, and other plant sources; used in the manufacture of *paper, rayon, plastics*, and *explosives.*

Cellulose Ethanoate. Cellulose acetate. An *ester* obtained by the action of *ethanoic anhydride* and *cellulose*. It is a white *solid*, used in the manufacture of *rayon* and *plastics.*

Cellulose Nitrate. Nitrocellulose. *Nitric acid ester* of *cellulose*. A range of *compounds* obtained by treatment of cellulose with a mixture of nitric and *sulphuric acids*; It is used in the manufacture of *plastics,* lacquers and *explosives.*

Cement. Any bonding material.

Centrifugal. Tending away from the centre;

Centrifugal Force. Resistance of revolving body, the virtue of its inertia, to acceleration towards the centre. It is equal and opposite to constraining force.

Centigrade (Celsius) Thermometer. Thermometer constructed by Celsius in which boiling point of water is 100° and freezing point zero degree.

Centripetal. Tending towards the centre ;

Ceraceous (Cereous). Waxy.

Cerium. Ce. Element. R.a.m. 140.12. At. No. 58 A steel-grey soft *metal;* It occurs in several rare minerals, e.g. *monazite* sand, and is used in *pyrophoric alloys* for lighter 'flints'; Its *compounds* are used in the manufacture of *gas mantles* and in *glass* polishing.

Cerium dioxide. Ceria. CeO_2. A white crystalline powder, m.p. $2600^{o}C$, Which is used in *glass* polishing.

Cermet. Ceramet. Abbreviation of CER (A) mic and metal. A very hard mixture of a *ceramic* substance and *sintered* metal, used where resistance to high *temperature, corrosion*, and abrasion is required.

Cerous. Containing the cerium (III) *ion.*

Cetane ($C_{16}H_{34}$). Hexadecane colourless solid hydrocarbon; It is used as reference fuel in determining ignition quality of automobile fuels.

Cetane Number. A measure of the ignition characteristics of a diesel fuel by comparison with a range of mixtures, in which *cetane* is given a arbitrary value of 100 and α -methylnaphthalene is 0.

Cetyl Alcohol. *Hexadecanol.*

C.G.S. System. Centimeter-gram-second system of physical *units* derived from the centimetre, *gram* mass and the *second.* It has been replaced for scientific purposes by the *SI units.*

Chabasite. A natural *zeolite,* calcium aluminium silicate.

Chalcedony. A veriety of natural impure *silica*, SiO_2. It has a fibrous structure and a waxy lustre, used for ornaments.

Chalcocite. Copper glance. Natural copper sulphide, Cu_2S. It occurs with other copper ores.

Chalcogens. Refers to the elements of group VIA of a *periodic table:* oxygen, sulphur, selenium, tellurium and polonium.

Chalcopyrite. Copper pyrites. A natural sulphide of copper and iron, (Cu, Fe)S_2; the most abundant ore of copper.

Chalk. Natural *calcium carbonate*, $CaCO_3$, formed from the shells of minute marine *organisms*. Blackboard chalk sticks are *calcium sulphate*, $CaSO_4$.

Chalones. Physiologically active substances produced within *tissues* that appear to control the *mitosis* of the specific tissues that produce them.

Chalybeate. Chalybite. Natural $FeCO_3$.

Charcoal. Amorphous form of carbon obtained by heating wood or other organic materials in absence of air.

Chemical Affinity. Tendency to combine with one another.

Chemical Bond. Force that keeps atoms in any molecule together.

Chemical Energy. Part of energy being stored in atom or molecule, can be released by chemical reaction.

Chemical Equation. Symbolic representation expressing true chemical reaction of various substances.

Chemical Formulae. Symbolic representation of molecule indicating its chemical composition *e.g.*, H_2O stands for water.

Chemical Inhibitor. Substance that can retard or stop a chemical reaction.

Chemical Kinetics. Branch of physical chemistry which deals with study of mechanisms and rates of chemical reactions.

Chemical Reaction. Interaction of two or more substances to form one or more new substances.

Chemical Warfare. Application of chemical energy for military purposes in contrast to physical energy as represented by impact of bullets or shock of explosives.

Chemiluminescence. Emission of light accompanied by some heat during a chemical reaction without producing any apparent change in temperature.

Chemistry. Study of composition and properties of elements and compounds including their combinations and action upon one another.

Chemotheraphy. The treatment of disease by chemical substances which are toxic to microorganism.

Chemurgy. The study of chemical industrial processes which are based on organic substances of agricultural origin.

Chert. A natural form of silica, SiO_2, resembling flint.

Chile Saltpetre. Impure *sodium nitrate*, $NaNO_3$. Found in huge deposits in Chile.

China Clay. Kaolin. A pure natural form of *hydrated* aluminium silicate, $Al_2Si_2O_5(OH)_4$. It loses *water* on being heated and changes its chemical composition. It is used for making *procelain.*

Chines White. *Zinc Oxide* ZnO.

Chip. **Silicon chip.** A crystal of a silicon *semiconductor* fabricated to carry out a number of electronic functions in an *integrated circuit.* Each chip is about one thousandth of a millimetre in diameter.

Chirality. The concept of 'handedness' (right- or left-handedness) applied to *stereoisomerism.* A geometrical figure representing the configuration of a *molecule* in space is said to have chirality if it cannot be made to coincide with its image in a plane mirror.

Chiron. A minor planet, that revolves around the Sun between the orbits of Saturn and Uranus.

Chitin. A complex organic substance, related to the *carbohydrates* but containing nitrogren. It forms and essential part of the shells of crustaceans and insects. It is also found in some fungi.

Chloracne. A skin disease that is caused by certain chlorinated aromatic hydrocarbons. It can result from contact, ingestion, or inhalation of the chemicals.

Chloral. *Trichloroethanal.*

Chloral hydrate. **2,2,2-*trichloroethanediol.***

Chloranil. $C_6C_{14}O_2$. A yellow *insoluble* crystalline substance, m.p. 290°C. It is used as a *fungicide* and in the manufacture of *dyes.*

Chlorargyrite. *Horn silver.*

Chlorate. A salt of a *chloric acid.*

Chloric Acids. Any of four *oxoacids*. The most common is chloric (V) acid, $HClO_3$. It is a strong *acid* and *oxidizing agent.*

Chloride A *salt* of *hydrochloric acid*, HCl.

Chloride of Lime. $CaOCl_2$. *Bleaching powder.*

Chlorination 1. The introduction of a chlorine atom into a compound by *substitution* or by an *addition reaction*. **2.** The treatment of drinking water with chlorine or a chlorine compound, such as *sodium hypochlorite* or *bleaching powder.*

Chlorine. **Cl.** Element. R.a.m. 35.453. At No. 17. A greenish-yellow poisonous gas, m.p. —100.98°C., b.p. —34.6°C., with a choking irritating smell. Manufactured almost entirely by the *electrolysis* of brine. Used in the manufacture of ***bleaching powder, disinfectants, hydrochloric acid*** and many organic compounds. Also used as a *germicide* in drinking water.

Chlorine Water. Aqueous solution of chlorine.

Chloroform ($CHCl_3$). Trichloro-methane; colourless liquid obtained by heating ethanol or acetone with bleaching powder; It is used as anaesthetic and in some drugs.

Chlorometer. Apparatus used for measuring available chlorine in bleaching powder; etc.

Chloromycetin. A drug used against typhoid, cerebro-spinal meningitis etc.

Chlorophyll. Green colouring matter in plants, essential for photosynthesis.

Chloroquin. Drug used to supress malaria.

Cholesterol ($C_{27}H_{45}OH$). White waxy sterol (fat derivative) found in tissues of human body, performs number of vital function. It excessive production causes coronary thrombosis.

Choline. **$OH.C_2H_4N(CH_3)_3OH$.** An *organic base*. It is a constituent of some *fats* and of egg yolk. It is a member of the *vitamin* B complex.

Chondrite. A type of stony meteorite. It contains the small round masses of *olivine* or *pyroxene* known as chondrules.

Chord (math.). A straight line joining two points on a curve.

Chromatin. Nucleoprotein which forms part of chromosome.

Chromatopsia. Coloured vision.

Chromic Acid. It is orange-red in colour; and is used in bleaching and dyeing.

Chromite. Mineral, double oxide of chromium and iron.

Chromium. Hard, bluish-white metal, quite stable in air; used for chromeplating and in manufacture of steel.

Chromium Plating. The deposition of a thin resistant film of chromium *metal* by *electrolysis. Solution* of *chromic acid* is used as electrolyte.

Chromium Steel. Steel containing varying amounts of chromium, 8—25%; it is strong and used for tools, etc.

Chromogene. Micro-organism capable of producing colour under suitable conditions.

Chromophore. Any chemical group that causes compounds to have colours *e.g.*, azo group.

Chromophyll. Any pigment of plant.

Chromosphere. The layer of the *Sun's* atmosphere which surround the *photosphere*, which is visible during a total *eclipse*. The chromosphere is several thousand miles thick and temperature of 20,000 K.

Chromous. Containing chromium in its + 2 *oxidation state, e.g.*, chromous chloride, $CrCl_2$.

Chromyl. The *bivalent radical* CrO_2^{-2}, containing chromium in its + 6 *oxidation state, e.g.* in chromyl chloride. CrO_2Cl_2.

Chronograph. An accurate time-recording instrument.

Chronometer. An accurate clock, used on a ship in navigation.

Chronon. A hypothetical particle of time defined as the ratio of the diameter of an *electron* to the *speed of light*: *i.e.* the time taken for light to transverse an electron. Approximately 10^{-24} *second.*

Chronotorn. A device that measures the time taken between two events, by measuring the positions on a transmission line of pulses initiated by the events.

Chrysotile. *Serpentine.*

Ciment Fondu. *Bauxite cement.*

Cinchonidine. $C_{19}H_{22}N_2O$. A white crystalline *alkaloid*, m.p. 207.2°C., used as a substitute for *quinine*. One of its *isomers*, cinchonine, m.p. 265°C., is also used for this purpose.

Cineole. $C_{10}H_{18}O$. A colourless oily *liquid terpene*, has an odour of *camphor*. It is found in certain *essential oils* and used in perfumes and medicine.

Cingulum. Part of plant between stem and root.

Cinnabar. Natural mercuric sulphide (HgS); bright red crystalline solid is principle ore of mercury.

Cis. Geometrical arrangement in which similar groups are adjacent to each other.

Citrate. Salt of citric acid.

Citric acid ($C_6H_8O_7$). White crystalline acid present in various citrus fruits.

Citrol ($C_{10}H_{16}O$). Pale yellow oily liquid having pleasant lemonodour; derived from oils of lime, lemon and citronella; used as flavouring agents.

Citronellal. $C_9H_{17}CHO$. A colourless *liquid aldehyde* existing in several isomeric forms, b.p. 205-8°C., It has lemon-like odour. It is used as a flavouring in the manufacture of perfume.

Citronellol. $C_3H_{17}CH_2OH$. A colourless *liquid alcohol* existing in se-

seral *isomeric* forms, b.p. 110°C., used in the manufacture of perfumes.

Cladding (Phys.). The covering of a *fuel element* in a *nuclear reactor* by a thin layer of another *metal*, which prevent corrosion by the *coolant* and the escape of *fission products.*

Clark Cell. A *primary cell*, used as a standard of *E.M.F.* It consists of a zinc *amalgam anode* and a mercury *cathode*, both immersed in a *saturated solution* of *zinc sulphate.*

Classical Physics. Refers to. *Physics* prior to the *quantum theory* (or in some senses prior to the theory of *relativity*).

Clathrate Compounds. Chemical *compounds* formed not by the action or *valence bonds*, but by 'molecular imprisonment', the combined *molecules* being held together mechanically by virtue of their configuration in space.

Claude Process. A process for producing *liquid air*, based on the cooling that results from the adiabatic expansion of a gas that is performing external work.

Clays. Finely-divided *rock* materials whose component *minerals* are various *silicates*, mainly of *magnesium* and *aluminium.*

Cleavage. The manner of breaking of a crystalline substance.

Clone. A group of organisms that have arisen from a single individual by asexual reproduction. All members of the group are genetically identical to each other and to the parent organism. Plants propagated by cuttings are clones, as are organisms produced from *somatic* cells.

Cloud Chamber (phys.). Wilson cloud chamber. An apparatus for making the tracks of ionizing particles visible as a row of droplets.

Cloud Point. The temperature at which a *homogeneous liquid* becomes cloudy or turbid due to separation into two *phases*, when cooled under specified conditions.

Cobalt (Co). Silvery-white, hard transition metal. It is malleable, ductile and slightly magnetic; used in various alloys and manufacture of high speed cutting tools.

Coenzyme. Organic molecule which act as cofactor of some enzyme.

Cofactor. Non-protein component, essential for catalytic activity or some enzyme.

Coffey Still. An apparatus for the *fractional distillation of solutions of ethanol* obtained by *fermentation* on an industrial scale; the product is known as *rectified spirit.*

Cohesion. It is the force of attraction between like molecules.

Coke. Fuel obtained by heating coal in enclosed space, whereby its more volatile constituents are driven off.

Colchicine $C_{22}H_{25}NO_6$. A yellow crystalline *alkaloid*, m.p. 156°C, obtained from the autumn crocus, interferes with the process of *mitosis* in such a way that it causes a doubling of the number of *chromosomes* in a cell. It is used as an artificial method of obtaining new agricultural and horticultural varieties and in the treatment of gout.

Colcothar Rouge. Red iron oxide used as a pigment (Fe_2O_3).

Cold Light. Light emitted in luminescence that is accompanied by little or no infra-red radiation and causes no heating effect.

Collagen. Protein in fibrous connective tissue, readity converted into gelatine.

Colloid. Substance in solution or suspension having particles which are too large to pass through organic membrane.

Colloidal Metals. *Colloidal solutions* or *suspensions* of *metals*, the metal being distributed in the form of very small electrically charged particles. They are used in medicine.

Colorimeter. An apparatus used in colorimetric analysis for comparing intensities of colour.

Colour. Visual sensation resulting from impact of light of some particular wave-length or frequency or retina of eye.

Colour-blind. One who is unable to distinguish some colours from others.

Colza Oil. Rapeseed oil. Yellow oil. Used as an edible oil, illuminant, lubricant, and in the *quenching of steel.* Mineral 'colza oil' is a mixture of *paraff in hydrocarbons* having a boiling range of 250°—350°C.

Coma 1. The nebulous patch of *light* surrounding the *nucleus* of a *comet.* **2.** Defect of a *lens* or spherical *mirror*, it causes a blurred *comet*-like image.

Combination (math.). A selection of a specified number of different objects from some larger specified number.

Combustion. Burning. A *chemical reaction*, in which a substance combines with oxygen producing *heat, light* and *flame.*

Comet. A heavenly body, moving under the attraction of the Sun in an eccentric orbit. It consists of a hazy gaseous cloud having a brighter nucleus and a fainter tail. The nucleus is thought to consist of ice and dust particles.

Command Guidance. A method of missile or *rocket* guidance in which computed information is transmitted to the missile and causes it to follow a directed flight path.

Compound. Homogeneous mixture of two or more elements, its constituents cannot be separated by ordinary physical methods.

Compounds Lens. Combination of two or more lenses joined together; one lens forms enlarged image of object and the other lenses magnify the image formed by first lense.

Compound Microscope. Microscope having two or more lenses.

Computor. Instrument capable of performing mathematical calculations on numbers.

Concave. Curving inwards; thus, a concave *lens* is thinner at the centre than at the edges.

Concavo-convex. Denoting a *lens* that curves inwards on one side and outwards on the other side.

Concentrated. A *reagent* containing the minimum of *water* or other *solvent.*

Concentration. *c*. The quantity of a substance present in quantity of another substance. The concentration of *aqueous solutions* is usually expressed in *moles* per cubic *metre.*

Concentration Cell. A *primary cell* whose *E.M.F.* is on account of a difference in *concentration* between different parts of the *electrolyte.*

Concentric. Having the same centre.

Conchoidal Fracture. An irregular break with a curved face exhibiting concentric rings.

Concrete. A building material consisting of stone, *sand, cement* and *water*. Reinforced concrete has *steel* rods imbedded in it to increase its *tensile strength.*

Condensation (chem.). A chemical change in which two or more *molecules* react with the elimination of *water* or of some other simple substance e.g. *ethanoic anhydride*, $(CH_3CO)_2O$, may be regarded as a condensation product of *ethanoic acid*, CH_3COOH, a molecule of the anhydride being formed when two molecules of the acid combine with the elimination of one molecule of water.

Condensation of Vapour. The change of *vapour* into *liquid*, that takes place when the *pressure* of the vapour becomes equal to the *vapour pressure* of the liquid at that *temperature.*

Conduplicate. Vernation in which each leaf is folded lengthwise around the next leaf.

Congenital. Refers to a disease existing from birth.

Conjugate Points. Of a lens. Points on either side of the *lens*, such that an object placed at either will produce an image at the other.

Conjunction (astr.). A *planet* is said to be in superior conjunction when it is in a straight line with the *Sun* and the *Earth*; a planet with its orbit inside that of the Earth is in inferior conjunction when it is between the Sun and the Earth and in line with them.

Connective. It is the bundle of nerve fibres in which two ganglia are united on the same side of the body.

Conservation of Charge. The principle that electric charge can be neither created nor destroyed.

Conservation of Mass and Energy. The law states that in any system energy and mass cannot be created or destroyed.

Conservation of Momentum. The principle that the total *momentum* of two colliding bodies before impact is equal to their total momentum after impact.

Consolute Temperature. Critical solution temperature. The temperature at which two partially miscible liquids become completely miscible.

Conspecific. Individual belonging to the same species.

Constant (math., phys.). Any quantity that does not vary.

Constantan. An *alloy* of copper containing 10%-55% nickel; it is used in electrical equipment.

Constant Boiling Mixture. *Azeotropic mixture.*

Contact Angle for Solid-liquid Interface. The *angle* included between the tangent plane to the surface of a *liquid* and the tangent plane to the surface of a *solid* at any point along their line of contact.

Contact Potential Difference. It two dissimilar *metals*, *a* and *b*, are in contact then a *difference* exists between point *A*, just outside *conductor a*, and a point *B*, just outside conductor *b*. This is the contact potential difference of the two conductors.

Contact Process. An industrial process for the manufacture of *sulphuric acid*, H_2SO_4. *Sulphur dioxide*, SO_2, is made to combine with oxygen by passing it over a heated *catalyst*, usually platinum or *platinized asbestos*. The *sulphur trioxide*, SO_3 so formed is combined with *water* to give sulphuric acid.

Containment. Confinement. In a controlled *thermonuclear reaction*, the process of preventing the *plasma* from coming into contact with the walls of the containing vessel is referred to as containment or confinement.

Continental Drift. The theory that all the continents on the Earth's surface were once a single land mass, which subsequently split up. It is postulated that some 135 million years ago the single land mass, named Pangaea, split into the northerly Laurasia and the southerly Gondwanaland. 65 million years ago these two continents split again, Laurasis into North America and Europe-Asia, and Gondwanaland into South America and Africa.

Continuous Wave. CW. *Radio radar* transmissions generated continuously and not in short *pulses.*

Continuum. A continuous series of component parts passing into one another; *e.g.*, the three *space* dimensions and the time dimension are considered to form a four-dimensional continuum.

Contour Feather. Feather having stiff shaft and vane. It gives a definite shape to the body.

Contract Vacuole. It is a vacuole present in protozoa. It fills and empties regularly.

Control Grid. An electrode placed between the cathode and the *anode* in a *thermionic valve* for controlling the flow of electrons through the valve.

Control Rod. That part of the control system of *nuclear reactor* which directly affects the rate of reaction. Usually a rod or tube, and can be moved up or down of its *axis*, made of steel or aluminium containing boron, cadmium, or some other strong absorber of *neutrons.*

Convection. Transference of *heat* through a *liquid* or a *gas* by the actual movement of the *fluid.* Portions in contact with the source of heat become hotter, less dense, and rise; their place is taken by colder portions, thus setting up convection currents.

Convex. Curving outside.

Convexo-concave. Convex on one side and concave on the other.

Convexo-convex. Convex on both sides.

Copal. A natural *resin* obtained from certain trees. Used in varnishes.

Coplanar (math.). In the same *plane.*

Copper. Cu. Element. R.a.m. 63.54. At. No. 29. A red *metal,* m.p. 1084°C., b.p. 2582°C., r.d. 8.95; after silver, it is the best *conductor* of electricity. It is used for steam boilers, electrical wire and apparatus, *electrotyping*, and in *alloys, e.g., bronze, brass, speculum metal, gun metal, bell metal, Dutch metal, manganin, constantan, nickel silver, German silver,* etc.

Copper Chlorides. $CuCl_2$, is a brownish powder, m.p. 620°C., obtained by action of chlorine over heated copper. The dihydrate is blue-green. Copper (I) chloride (cuprous chloride), CuCl, m.p. 430°C., is obtained by boiling copper (II) chloride with copper and *hydrochloric acid.* It is used as a *catalyst.*

Copper Oxides. Copper (I) oxide (cuprous oxide or red copper oxide), Cu_2O, is a red *insoluble* powder, m.p. 1235°C., which occurs naturally as *cuprite*. It is use as a *fungicide*. Copper (II) oxide CuO, is a black insoluble substance, m.p. 1326°C.

Copper (II) Sulphate. Cupric sulphate, blue vitriol. $CuSO_4 . 5H_2O$. A blue crystalline *soluble salt*, obtained by action of copper (II) oxide with dilute sulphuric acid. Used as a *mordant*, and *fungicide*. It is also used in *electroplating* and preservation of wood.

Cordite. An *explosive* obtained from *cellulose nitrate* and *nitroglycerine.*

Core. 1. A magnetic material used to increase the inductance of a coil through which it passes. **2.** The central part of a *nuclear reactor* containing the fissile material. **3**. The devices, that constitute the memory of a computer.

Coriolis Force. A fictitous force used to simplify calculations involving rotating systems, such as the movement of air on the surface of the Earth.

Cornea. Outer transparent tissue of eye that admits light in eye.

Corona. A white irregular halo surrounding the *Sun*, is visible during a total *eclipse.*

Corona Discharge. A luminous discharge that appears round the surface of a *conductor* on account of *ionization* of the air (or other *gas* surrounding it). It is caused by the *voltage* gradient exceeding a critical value, but not being sufficient to cause sparking.

Corpuscular Theory. The theory according to which *light* consists of minute corpuscles in rapid motion.

Corrosion. Surface chemical action, on *metals* by the action of moisture, air, etc.

Corrosive Sublimate. *Mercury (II) chloride.* ($HgCl_2$).

Corticosteroid. A *hormone* produced by the cortex of the adrenal glands. These are used as anti-inflammatory drugs and regulate the body's use of *carbohydrates, fats,* and *proteins,* and mineralccor-ticoids which regulate the *salt* and *water* balance.

Cortisone. 17-hydroxy-11-dehydrocorticosterone. $C_{21}H_{28}O_5$. A crystalline *corticosteroid hormone*, m.p. 215°C., secreted by the cortex of the adrenal gland. It reduces local inflammation, is used in the treatment of rheumatic and other conditions.

Corubin. Crystalline *aluminium oxide*, Al_2O_3, obtained as a *by-product* in the *aluminothermic reduction.*

Corundum. Natural *aluminium oxide*, Al_2O_3. A crystalline substance nearly as hard as *diamond*, used as an *abrasive.*

Cosmic Rays. Very energetic *radiation* falling upon the Earth from outer *space*, consisting chiefly, of charged particles. The majority of these are *protons*, although *electrons* and *alpha particles* are also present. High-energy electrons are created in the atmosphere by meson *decay*, by interaction of high energy protons with nuclei, by *knockon* collisions of mesons of mesons with electrons, etc. These high-energy electrons give rise to *cosmic ray showers* resulting in the creation of *photons, positrons*, and further electrons. Energies as high as 10^{20} *electronvolts* have been observed with cosmic ray particles.

Cosmogony. The science of the nature of the heavenly bodies, with particular reference to the formation of *planets, stars* and *galaxies.*

Cosmology. The science of the nature, origin, and history of the *Universe.*

Consmotron. A *proton accelerator* that contains a very large ring-shaped *electromagnet.*

Cottrell Precipitator. A device used for removal of dust particles from gases by *electrostatic precipitation.*

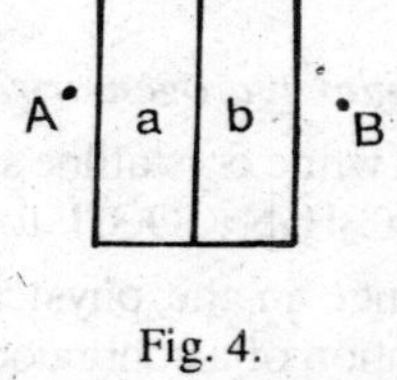

Fig. 4.

Coude System. A form of astronomical reflecting *telescope* in which *light* from the primary *mirror* is reflected back along the axis of the telescope by means of a system of mirrors.

Coulomb. The *SI unit* of *electric charge*, defined as the quantity of electricity transferred by 1 *ampere* in one second. 10^{-1} *electromagnetic unit*; 3×10^9 *electrostatic units*. Symbol C.

Coulomb Scattering. The *scattering* of sub-atomic particles caused by the *electrostatic* (coulomb) *field* surrounding an atomic *nucleus.*

Coulomb's Law. The *force* of attraction or repulsion between two charged bodies (assuming that the charges behave as though they were concentrated at a point), is proportional to the magnitude of the charges and inversely proportional to the square of the distance between them.

Coumarin. $C_9H_6O_2$. A white crystalline substance, m.p. 71°C., It has an odour of vanilla. Used as a flavour and in perfume.

Coumarone. Benzofuran. $C_6H_4OCH{:}CH$. A *liquid* obtained from *coal-tar*, b.p. 173°C, it polymerizes into a synthetic *resin*; used in the *paint* and varnish industry.

Counter Tube. A device for counting individual ionizing events.

Couple (phys.). Two equal and opposite parallel, but not colinear, *forces* acting upon a body. The *moment* of a couple is the product of either force and the perpendicular between the line of action of the forces.

Covalent Crystal. A crystal in which the *atoms* are held in the *lattice* by covalent bonds. Typical examples are *diamond* and silicon (IV) oxide.

C.P.U. Central processing unit. The central *electronic* unit in a *computer* that processes input information, and information from the *store*, and produces the output information.

Craking (chem.). Pyrolysis. The *decomposition* of a chemical substance by action of *heat*; the conversion of *mineral oils* of high *boiling point* into more *volatile* oils suitable for petrol engines, by 'cracking' the larger *molecules* of the heavy oils into smaller ones. In catalytic cracking the decomposition takes place in the presence of a *catalyst.*

Cream of Tartar. *Potassium hydrogentartrate.* $C_4O_6H_5K$.

Creatinine. $C_4H_7N_3O$. A white crystalline substance, derived from the *amino acid* creatine, $C_3H_8N_3COOH$, found in urine, blood, etc.

Creep. A permanent change in the physical dimensions of a *metal* caused by the application of a continuous *stress.*

Creosote. A *distillation* product obtained from *coal-tar*. An oily, transparent *liquid* containing *phenol*, it is used for preserving timber.

Cresol. *Methylphenol.*

Cristobalite. Mineral consisting of silicon (IV) oxide, SiO_2.

Crith. The *mass* of 1 litre of hydrogen at 0°C, and a pressure of 760 mm; approximately 0.09 g.

Critical Angle. Of a medium (phys.) the least angle of *incidence* at which *total internal reflection* occurs.

Critical Damping. A measuring instrument is said to be critically damped if it takes up its *equilibrium* deflection in the shortest possible time, the oscillations of the indicator (needle) about the equilibrium position being quickly damped out. *Galvanometers* are normally used critically damped.

Critical Mass. The minimum *mass* of *fissile* material required in a *nuclear reactor* or a *nuclear weapon* to sustain a *chain reaction.*

Critical Potential. Refers to the minimum *energy* required to raise the *energy level* of an *orbital electron* or to remove it from the *atom.*

Critical Pressure. The pressure of the *saturated vapour* of a substance at the *critical temperature.*

Critical State. Refers to the state of a substance when its liquid and gaseous *phases* have the same *density*, at the same *temperature* and *pressure.*

Critical Temperature. Means the *temperature* above which a gas cannot be liquefied by application of *pressure* alone.

Critical Velocity. Means the velocity at which the flow of a *liquid* cease to be *streamline* and becomes *turbulent.*

Critical Volume. The *volume* occupied by unit mass of a substance at its *critical temperature* and *critical pressure.*

Cross-linkage (chem.). The joining of polymer *molecules* to each other by *valence bonds.*

Cross-section. In *nuclear physics* the cross-section represents the effective area that has to be attributed to a particular *atom* or *nucleus* to account geometrically for its interaction with an incident *beam* of *radiation.*

Crotonic Acid. Butenoic acid. $CH_3CH{:}CHCOOH$. A colourless crystalline *soluble* substance that exists in two *isomeric* forms. *Trans*-but-2-enoic acid has a.m.p. of 71.6°C. and is used in organic synthesis. *Cis*-but-2-enoic acid (isocronic acid) has a m.p. of 145°C.

Crown Glass. A variety of *glass* having potassium or barium it is less fusible than ordinary *soda glass*, is used in optical instruments.

Crucible. A vessel of heat-resisting material used for containing high temperature *chemical reactions.*

Cryogenics. The study of materials and phenomena at *temperatures* close to *absolute zero.*

Cryohdrates. Crystalline substances, containing the *solute* with a definite molecular proportion of *water*, that crystallize out from *solutions* cooled below the *freezing* point of pure water.

Cryolite. Natural sodium aluminium fluoride, Na_3AlF_6. Used in the manufacture of *aluminium.*

Cryometer. A *thermometer* especially designed to measure low *temperatures.*

Cryophorus. An apparatus used to demonstrate the cooling effect of *evaporation.*

Cryoscopic Method. Freezing-point method. The determination of the *relative molecular mass* of a dissolved substance by noting the *depression of freezing point* produced by a known *concentration* of it in *solvent.*

Cryostat. A vessel in which a specified low *temperature* may be maintained.

Cryotron. A switch based on *superconductivity.* The simplest form consists of a coil of wire of one superconducting material wound round a length of wire of another superconductor, all immersed in a bath of liquid helium.

Crystal. A solid substance that has definite geometrical form. Most solid substances, when pure, have a definite crystalline form. Solids that do not form crystals are said to be *amorphous.* Crystals are classified on the basis of their *lattices*, or according to the type of *bond* that holds them together, *i.e., electrovalent* (or *ionic crystals*), *covalent crystals*, or *metallic crystals.*

Crystal Counter. A *counter tube* which depends upon a *crystal* in which the electrical *conductivity* is momentarily increased by an ionizing event.

Crystal Detector. A fine wire in contact with a crystal of *galena* (PbS) or other suitable *semiconductor.* This arrangement acts as a good conductor of *electricity* in one direction, and suppresses the flow of electricity in the other direction.

Crystalline. Consisting of *crystals* or having a regular structure of *atoms, ions,* or *molecules* characteristic of crystals.

Crystallite. A small *crystal*, as in a microcrystalline substance.

Crystallization. The formation of *crystals* in a *liquid* or *gas.*

Crystallograph. The study of the geometrical form of *crystals.*

Crystalloids. Substances that, in *solution*, are able to pass through a *semipermeable membrane*; substances which do not usually form *colloidal solutions.*

Crystal Microphone. A microphone in which the sound waves to be amplified or transmitted vibrate a *piezoelectric crystal,* which generates a varying E.M.F.

Crystal Oscillator. A source of electrical oscillations of very constant *frequency* determined by the physical characteristics of a *piezo-electric crystal.* The crystal, usually *quartz,* it either produces the oscillation in a tuned circuit or is coupled to a tuned circuit to control its frequency.

Crystal Pick-up. A pick-up in a record player in which the varying *E.M.F.* in produced by *piezoelectric crystal* due to the vibrations obtained from the undulations in the groove of the record.

Crystal Rectifier. A *semiconductor diode* used as a *rectifier* in a manner similar to a *diode valve.*

Crystal Systems. The seven classes into which crystals are divided : *cubic* tetragonal, orthorhombic, hexagonal, trigonal, monoclinic and triclinic.

CS $C_6H_4ClCH:C(CN)_2$. A white powder, m.p. 52°C., used as a tear gas and 'harassing agent' in crowd control.

Cube 1. A regular hexahedron; a regular solid figure with six square faces.

Cube Root $\sqrt[3]{}$. The cube root of a number is the quantity which when raised to the third power gives that number. Thus 2 is the cube root of 8.

Cubic Centimetre Cm^3 or cc. A metric unit of volume sometimes used synoymously with *millilitre,* ml., one thousandth of a litre.

Cubit Crystal. A *crystal system* in which the *unit cell* is a *cube.* In the simple cubic system there are *lattice point* at the cube's eight corners.

Culture medium. A nutrient preparation used to grow and cultivate *microorganisms* for experimental purposes.

Cumene. (1-methylethyle) benzene. A colourless liquid *aromatic hydro- carbon,* b.p. 152°C. Found in *petroleum* and is used as an *intermediate* in *organic synthesis.*

Cupellation. The separation of silver, gold, and other *noble metals* from impurities which are oxidized by hot air. The impure metal is placed in a flat dish made of porous *refractory* material known as cupel and a blast of hot air is directed upon it in a special furnace. The impurities are oxidized by the air and are partly swept away by the blast and partly absorbed by the cupel.

Cuprammonium Ion. The bivalent tetraamminecopper (II) cation, $[Cu(NH_3)_4]^{2+}$, formed by copper with *ammonia ligands.*

Cupric. Containing copper in its + 2 *oxidation state*, *e.g*,. cupric chloride, $CuCl_2$.

Cuprite. *Copper oxides.*

Cupro-nickel. An *alloy* of copper and nickel used in coinage.

Cuprous. Containing copper in its + 1 *oxidation state* e.g. cuprous. chloride, CuCl.

Cuprous Iodide (Cul I). White crystalline powder, prepared by adding potassium iodide solution to copper sulphate solution, it is insoluble in water and is not oxidised by air.

Curie. Unit of measure of specific activity of radioactive preparation.

Curic Point. Temperature of a given ferromagnetic substance above which it becomes paramagnetic; also called curie temperature.

Curie's Law. Magnetic susceptibility of paramagnetic substance is inversely proportional to absolute temperature.

Curie Therepy. Treatment of disease by radium.

Curium Cm. *Transuranic element*, At. No. 96. A radioactive *actinide.* Its most stable *isotope*, curium-247, has a *half-life* of 1.6×10^7 years.

Current Balance. An instrument for the determination of an *electric current* by measuing the *force* that the current produces between conductors.

Current Density. **1.** The current flowing through a conductor, *plasma*, etc., per unit cross-sectional area. It is usually expressed in amperes per square meter. **2.** (in *electrolysis*) The current flowing per unit area of electrode.

Cursor. A transparent slider on a measuring instrument or an indicating device, such as a spot of light.

Cyanic Acid. Acid composed of hydrogen, cyanogen and oxygen (HCNO).

Cyanaogas. Fine, black powder containing calcium cyanide, when comes in contact with moisture, it liberates hydro-cyanic acid (HCN) gas. It is used as fumigant in rat burrows.

Cyanogas Chloride (CICN). Colourless, poisonous gas, dissolves in alcohol, their and water.

Cyanogen [$(CN)_2$]. Colourless, poisonous gas, very soluble in water combustible and burns with violet flame to give carbon dioxide and nitrogen; essential ingredient of Prussian blue.

Cyanometer. Instrument used to measure blueness of ocean or sky.

Cycle. Set of changes or operations that a system undergoes finally coming back to its original position *e.g.*, nitrogen cyle.

Cyclization. Reaction in which straight chain compound is converted into cyclic compound.

Cyclo-hexane(C_6H_{12}). Colourless liquid obtained on reduction of benzene with hydrogen in presence of nickel at 200°C; used as solvent.

Cyclohexanole ($C_6H_{11}OH$). Crystalline soluble substance boils at 161°C; used as solvent.

Cyclo-pentane (C_5H_{10}). Colourless liquid obtained from petroleum boiling point is 50°C and is used as solvent.

Cyclo-propane (C_3H_6). Colourless gas with sweetish smell; used as an anaesthetic.

Cyanamide (NH_2CN). A colourless crystalline *unstable* substance, m.p. 44°C.

Cyanate, Fulminate. A *salt* or *ester* of *cyanic acid* (fulminic acid).

Cyanic Acid HOCN. An explosive *liquid*, also known as fulminic acid. Its *salts*, the *cyanates* (or fulminates) are also explosive.

Cyanide. A *salt* of *hydrocyanic acid*, HCN. Cyanides are intensely poisonous.

Cyanid Process. The extraction of gold from its *ores* by dissolving the gold in a *solution* of potassium cyanide, KCN, reducing the resulting potassium aurocyanide, $KAu(CN)_2$, with zinc, filtering off, melting down, and cupelling the metal.

Cyanine Dyes. A group of *dyes* containing a chain of carbon atoms with *conjugated double bonds* forming a bridge between two *heterocyclic nuclei*. Used for photographic *sensitization.*

Cyanite (Al_2SiO_5). A blue mineral consisting of aluminium silicate, used as a *refractory* material.

Cyanocobalamin Vitamin (B_{12}). $C_{63}H_{90}O_{14}N_{14}PCo$. A red crystalline substance, obtained from liver, eggs, fish etc., used in the treatment of pernicious anaemia, and to promote the growth of livestock.

Cyanogen (C_2N_2). A colourless, poisonous *gas* having a smell of bitter almonds. In chemical properties it resembles the *halogens*, forming cyanides analogous to the *chlorides*. It can be obtained by heating mercury (II) cyanide.

Cyano Group. The *univalent radical* —CN.

Cyanoguanidine. *Dicyandiamide.*

Cyanuric Acid. Tricyanic acid. $C_3H_3O_3N_3.2H_2O$. A white crystalline substance, used in organic synthesis consist of a six-membered ring of altern ting—NH—and—CO—groups.

Cybernetics. Refers to the theory of communication and control mechanisms in living beings and machines.

Cybotaxis. The tendency of liquid molecules to form regularly arranged groups, resembling *crystals.*

Cycle (phys.). Any series of changes or operations performed by or on a system, which brings it back to its original state, e.g. the *frequency* of an *alternating current* is measured in cycles per second.

Cyclic (chem.). Having ring structure.

Cyclic Figure (math.). A figure through all the vertices or corners of which a *circle* may be drawn. A figure inscribed in a circle.

Cyclic Quadrilateral. A four-sided plane *rectilinear* figure through the vertices of which a circle may be drawn. In such figure the pairs of opposite angles are *supplementary (i.e.*, total 180°).

Cyclization. The process of conversion of an open chain molecule into a *cyclic* compound.

Cycloalkane. A *saturated cyclic* compound having the general formula C_nH_{2n}. They are chemically similar to the *alkanes* but less reactive.

Cyclohexane C_6H_{12}. A colourless flammable *liquid*, b.p. 81°C, consisting of a six-membered ring. It is used as a *solvent* and in the manufacture of *plastics.*

Cyclohexanol ($C_6H_{11}OH$). Crystalline substance, m.p. 25.1°C, b.p. 161.1°C., used as a *solvent.*

Cycloid. Figure traced out in space by a point on the circumference of a circle, which rolls without slipping along a fixed straight line.

Cyclonite. Hexogen, R.D,.X. $(CH_2N.No_2)_3$. A powerful *explosive* made from *hexamine.*

Cyclopentane (C_5H_{10}). A colourless *liquid* b.p. 49.2°C, obtained from *petroleum* and used as a *solvent.*

Cyclopropane (C_3H_6). A colourless flammable *gas*, used as an *anaesthetic.*

Cyclotron. An *accelerator* for imparting to charged particles of atomic magnitudes *energies* of several million *electron-volts*. The *ions* or charged particles are caused to transverse a spiral path between two hollow *semicircular electrodes*, called dees, by means of a suitable *magnetic field* applied perpendicularly to the plane of the dees. At each half-revolution the particles receive an energy increase of some tens of thousands of electronvolts from an oscillating *voltage* applied between the dees.

Cylinder. A solid figure traced out by a rectangle round one side as axis. For a cylinder having vertical height h and radius of base r, the volume is $\pi r^2 h$ and the total surface area $2\pi r(h+r)$.

Cysteine. Crystalline amino acid found in most proteins.

Cystine. Sulphar-containing amino acid insoluble in water; involved in metabolism.

Cystocarp. Swollen urn-shaped fruiting body found in some red-algae.

Cystochemistry. Chemistry of living cells.

Cytokinin (s). Group of plant harmones that help to promote cell division in plants.

Cytosine. Pyrimidine of nucleic acid, isolated on acid hydrolysis of nucleic acid.

Cytosol. Soluble fraction of cytoplasm left after all particles have been removed by centrifugation.

D

DAC. Abbreviation for digital to analog converter.

D/A Converter. Abbreviation for digital to analog converter. A device for converting a digital signal into an equivalent analog signal.

Dactyl. Finger, digit.

Damp Air. Air having high relative humidity.

Damping Off. Disease of seedlings caused by soil-borne fungi, encouraged by cold, soil moisture and crowded conditions.

Daniel cell. Symbolic representation of cell; outer copper vessel containing concentrated solution of cupric sulphate and copper plate (positive electrode), inside is a porous pot containing zinc rod (negative electrode) dipped in dilute solution of zinc sulphate.

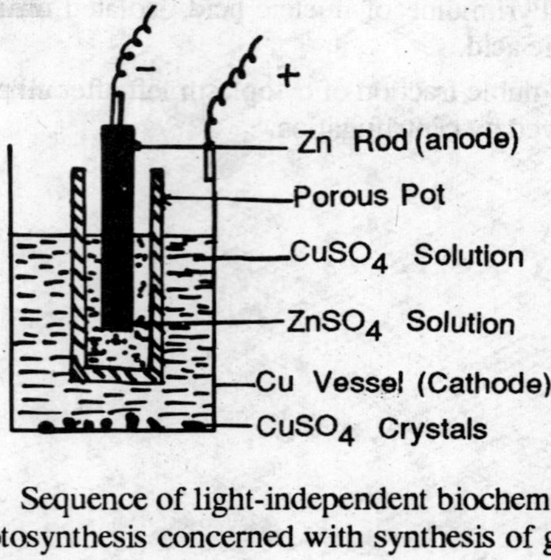

Dark Reaction. Sequence of light-independent biochemical reactions during photosynthesis concerned with synthesis of gluclose from carbon dioxide.

Dash-pot. A mechanical damping device, depends upon the fact that when a body moves through a *fluid* medium, viscous *forces* are set up, which damp the motion of the body. It usually consists of a piston, attached to the part whose movements are to be damped, fitted loosely into a cylinder containing either air on oil.

Dasymeter. A devise consisting of thin glass globe, used for determining density of gas.

Dasyphyllous. Having crowded, thick or wooly leaves.

Database. A collection of data stored in a *computer* and coded in such a manner that it can be extrated in various ways, under various headings.

Dating. The determination of the age of *mineral, fossil*, or wooden objects by measuring their *radioactivity.*

Dative Bond. Coordinate bond. A *covalent type* of *bond* in which both *electrons* forming the bond are donated by one *atom.*

Davy's Lamp. Safety-lamp for coal-miners.

Day-vision. Vision present in bright day light.

DDT ($C_{14}H_9CI_5$). Dichloro-diphenyl trichloroethane. It is white amorphous powder, insoluble in water, highly soluble in aromatic hydrocarbons. It is contact and stomach poison having longer residual action, used as an insecticide.

Deactivation. Loss of radioactivity.

Deaeration. Removal of air or any gas from a substance.

Deaminase. Enzyme that catalyses removal of amino group from a compound.

Deamination. Removal of amino group from a compound.

Dean and Stark Method. A method of estimating the quantity of *water* in an *oil* or other *liquid* substance.

De Broglie Wavelength. For a particle of *mass m* moving with *velocity v* the *wavelength* of the associated de Broglie wave is given by $\lambda=h/mv$, where *h* is *Planck's constant.*

Debye. A unit of molecular *dipole* moment equal to 1×10^{-18} *electrostatic unit* or 3.33564×10^{-39} *coulomb metre.*

Deca. Prefix denoting ten times in the *metric system*. Symbol da, *e.g.*, dam = 10 *metres.*

Decane ($C_{10}H_{22}$). Liquid paraffin hydrocarbon.

Decantation. Process used for separation of solid from liquid by allowing the former to settle down and pouring off the latter.

De-carboxylation. Process of removal of carbon dioxide from carboxyl group of organic acid.

Decay. 1. The *transformation* of a radioactive substance into its daughter products. **2**. The transformation of a particle into a more stable particle.

Decay constant. *Disintegration constant.*

Deci-. Prefix denoting one tenth in the *metric system*. Symbol d, *e.g.*, dm = 0.1 *metre.*

Decibel. One tenth of a *bel*. A unit for comparing levels of *power*. Two power levels, P_1 and P_2, are said to differ by *n* decibels when:

$$n = 10 \log_{10} P_2/P_1$$

Declination (astr.) The *angular distance* of a heavenly body from the *celestial equator.*

Decomposer. Organism that feeds on dead organic matter.

Decomposition. Decay. Breaking up of chemcial compounds into simpler molecule.

Decompound. Compound leaf with leaflets made up of distinct parts.

Decrepitation. The bursting or cracking of *crystals* of certain substances on heating. It is due to expansion of water within the crystals.

Defect. A discontinuity in the pattern of *atoms, ions*, or *electrons* in a *crystal*. A 'point defect' consists of *vacancy* or an *interstitial.* A 'line defect' is caused by a *dislocation.*

Deficiency Diseases. Diseases caused by lack of a particular *vitamin, amino acid mineral,* or other essential food factor in the diet; *e.g.*, scurvy, caused by the deficiency of vitamin C.

Defoliant. Chemical which induces premature leaf-fall.

Deformation. Any alteration in the size or shape of a body.

Deformation Potential. The *electric potential* that acts on a *free electron* in a *conductor* or *semiconductor* as a result of *deformation* of the *crystal lattice.*

Dehydrate. To remove water.

Denydrogenase: Any oxidising enzyme, it catalyses oxidation of substrate by removing hydrogen from it.

Depth of Field. The range over which a *camera*, or other optical instrument, will produce a distinct *image* of an object.

Derivative (chem.). A *compound* derived from (but not necessarily prepared from) some other compound, usually retaining the general structure of the parent compound; e.g. *nitrobenzene*, $C_6H_5NO_2$, is a derivative of *benzene.*

Derivative (math.). Derived function. The result of *differentiation* of a mathematical function.

Desalination. It is the process of removal of *salt* from *sea-water* to make it suitable for agricultural purposes or drinking.

Desiccation. Drying : removal of moisture.

Desiccator. An apparatus used in laboratories for drying substances and for preventing *hygroscopic* substances from picking up moisture. It consists of a glass vessel, with a close-fitting ground lid, that contains some hygroscopic substance, e.g. $CaCl_2$.

Desorption. The removal of *molecules*, *ions*, etc., from the surface of a solid so that they become gaseous; the reverse of *adsorption.*

Destructive Distillation. Heating a complex substance to produce chemical changes in it, and distilling off the *volatile* substances so formed.

Detector. That part of a *radio* receiver in which the information is separated from the *modulated carrier wave*. It is now called a demodulator.

Detergents. Washing or cleaning agents, consisting of *surfactants* or mixtures of these with other agent for specific purposes.

Detinning. The recovery of metallic tin from scrap tin-plate by the action of chlorine.

Detonating Gas. A *mixture* of hydrogen and oxygen in a *volume* ratio of 2:1. It is extremely explosive when ignited.

Detonation. Extremely rapid *combustion* which takes place within a high velocity *shock wave.* Also used to describe the combustion reactions that occur during *knocking* or 'pinking' in an *internal-combustion engine.*

Detritus. Organic debris from decomposing animals and plants; unconsolidated sediments derived from pre-existing rocks by natural processes.

Dew. Minute droplets of water produced by condensation of vapour in air when temperature falls sufficiently for vapour to reach saturation.

Dew Point. Temperature at which dew formation starts.

Dew Point Hygrometer. Instrument for determining dew point. It is determined by measuring the temperature at which vapour being cooled in silver vessel begins to condences.

Dextran. Any polysaccharide yielding glucose units on hydrolysis.

Dextrin. Polysaccharide obtained by partial hydrolysis of starch.

Dextrose. ($C_6H_{12}O_6$). Glucose; Grape-sugar. Crystalline sweet substance; used as fruit preservative and in medicines as sweetening agent.

Di-. Two.

Dialysis. Process by which large molecules like proteins, starches etc., are separated from small molecules such as amino acids and salts. This is done by selective diffusion through a semipermeable membrane. Dialysis is performed naturally by kidneys to extract wastes from blood, it is done artificially by machine if kidneys are damaged.

Diamond. An allotropic form of carbon, hard substance (melting point 7350°C). Not soluble in any solvent and bad conductor of heat and electricity; used in jewellery and cutting glass.

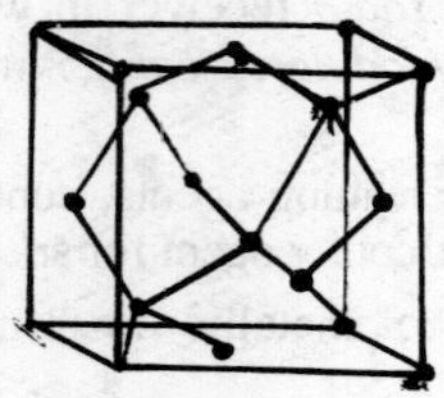

Fig. 5. The structure of diamond.

Diastase. Enzyme that converts starch into sugar. Present in germinating seeds and in pancreatic juice.

Diatom. Molecule containing two atoms — chlorine (Cl_2), hydrogen (H_2), nitrogen (N_2), oxygen (O_2) etc.

Diazomethane (N_2CH_2). Highly poisonous and explosive, yellow gas having a peculiar musty smell.

Diazonium Compounds. *Organic compounds* of the general formula $RN_{2+}X^-$, where R is an aryl radical, RN_2^+ is a cation, and X^- is an *anion, e.g.*, benzenediazonium chloride $C_6H_5N_2^+Cl^-$.

Dibasic Acid. An *acid* containing two *atoms* of replaceable *hydrogen* in a *molecule*; an acid giving rise to two series of *salts*; *e.g.*, *sulphuric acid*, H_2SO_4, gives rise to normal *sulphates* and hydrogen-sulphates.

Di (benzenecarbonyl) Peroxide. Benzoyl peroxide. $(C_6H_5CO)_2O_2$. An insoluble crystalline explosive solid, m.p. 106 — 8°C., used in bleaching flour, fats, oils, etc.

Dibromoethane. Ethylene dibromide. $C_2H_4Br_2$. A *voalatile liquid* has two *isomeric* forms. The common isomer 1, 2-dibromoethane, m.p. 10°C., b.p. 131°C., is used with anti-knock compounds in *petrol* and as a *solvent.*

Dibutyl Oxalate $(C_4H_9OOC)_2$. A colourless *liquid*, b.p. 243.4°C., used as a *solvent*.

Dichlorodifluoromethane CC_2F_2. A colourless *gas*, b.p. — 29°C., used as a *refrigerant* and as a *propellant* for *aerosols*.

Dichloroethane. Ethylene dichloride, Dutch liquid. $C_2H_4Cl_2$. An oily *liquid* has two *isomeric* forms. The common isomer 1, 2-dichloroethane, m.p.— 35°C., b.p. 84°C., is used as a *solvent* and in the manufacture of *polyvinyl chloride*.

Dichloromethane. Methylene chloride. CH_2Cl_2. A colourless *volatile liquid*, b.p. 40.1°C., used as a *solvent*, a *refrigerant*, and an *anaesthetic*.

Dichroism. Property of showing different colours shown by double refracting crystals when viewed in different directions by transmitted light.

Dichromate (bicromate). Salt of dicromic acid ($H_2Cr_2O_7$) having two chromium atoms.

Dichromatic. Having two colours in different individual of same species.

Dichroscope. Instrument used for testing dichroism of crystals.

Dielectric. Non-conducting; transmitting electric effects without conducting.

Diethyl. Having two ethyl groups.

Dimegaly. Having sperms of two different sizes.

Dimer. Polymer of the components.

Dimethyl Ether [$CH_3)_2O$]. Gas, soluble in water; used as refrigerating agent and low temperature solvent.

Dimorphism. Difference in size, structure, form, colour etc., between two types of individuals of same species.

Dinitrogen (N_2O). Colourless gas prepared by heating ammonium nitrate upto 270°C; used as anaesthetic; also called laughing gas.

Diode. A *therminoic valve* containing two *electrodes, anode* and *cathode*. The diod is used chiefly for *rectification* and *demodulation*. Modern diodes consist of a P-N *semiconductor junction*.

Diol (Dihydric Alcohol). Alcohol containing two hydroxyl groups per molecule of compound.

Diplo-. Double.

Diplopia. Double vision; seeing one object as two.

Dipolar. Having two poles;

Disperse Dyes. Dyes of all chemical types which are applied in the form of fine suspensions in water to man-made fibres, such as *cellulose acetate*, *nylon*, and *polyester* fibres. They are insoluble in water, but are soluble in organic solvents.

Disperse Phase. The disolved or suspended substance in a *colloidal solution.*

Dispersion (chem.). A system of particles dispersed and suspended in a *solid, liquid*, or gas.

Dissimilation. Process of physiological burning or breaking down, resulting in release of energy.

Dissipiate. Dispel; scatter; squander; separate and disappear.

Dissoluble. Capable of being dissolved.

Dissolve. To make solid or gas disappear in a liquid.

Distillate. Liquid that is vapourished and then condensed in process of distillation.

Distillation. Process of converting liquid into vapour and condensing vapour to liquid.

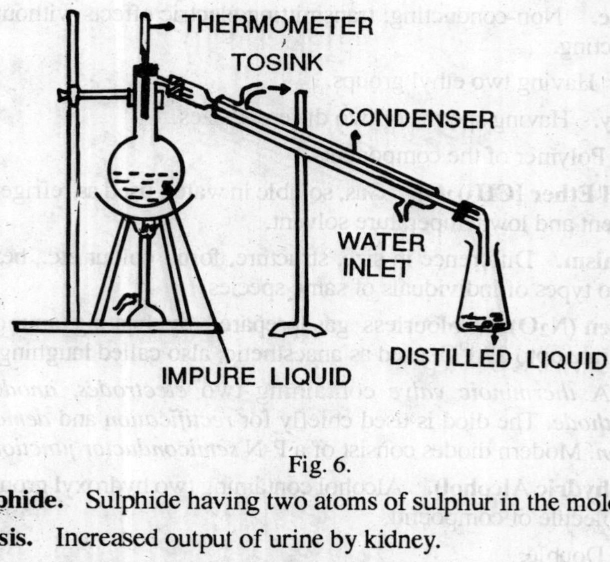

Fig. 6.

Disulphide. Sulphide having two atoms of sulphur in the molecule.

Diuresis. Increased output of urine by kidney.

Diuretic. Compound that increases activity of kidney.

Diurnal. Active during day.

Divalent. Bivalent. Having a *valence* of two.

Divergent. Going away in different directions from a common path or point.

Diverging Lens. A *lens* that causes a parallel *beam* of *light* passing through it to diverge or spread out.

Divers' Liquid. A solution of *ammonium nitrate* in liquid *ammonia.* It is used as a *solvent* for some *metals* and their *oxides* and *hydroxides.*

Divinyl Ether. Vinyl ether. $H_2C{:}CHOCH{:}CH_2$. A colourless inflammable *liquid* b.p. 39°C., used as *anaesthetic.*

Division. A arithmetic operation in which a dividend is divided by a divisor to give a quotient and a remainder.

DNA. *Deoxyribonucleic acid.*

Docosanoic Acid. Behenic acid. $CH_3(CH_2)_{20}COOH$. A crystalline *saturated fatty acid*, m.p. 80°C., used in the manufacture of cosmetics and *waxes*, and as a *plasticizer.*

Dodecahedron. A *polyhedron* having twelve faces.

Dodecanoic Acid. Lauric acid. $CH_3(CH_2)_{10}COOH$. A white crystalline *insoluble* substance, m.p. 44°C., used in the manufacture of *soaps, detergents* and cosmetics.

Dodecanol. Lauryl alcohol. $CH_3(CH_2)_{11}OH$. A white crystalline *insoluble* substance, the commercial form of which consist of a mixture of *isomers* with m.p. in the range 20-30°C. used in the manufacture of *detergents.*

Dorsal. Refers to back or upper surface.

Dorsum. Refers to back of animal.

Double Bond. Two covalent bonds that link two atoms in a chemical compound.

Double Helix. Molecule having two similar polymeric chains coiled in same direction about the same axis.

Double Salt. Compound of two salts obtained by crystallisation from solution containing both the salts.

Double Star. Two *stars* held very close to each other due to their mutual gravitational attraction, they move through *space* together giving the appearance to the naked eye, of being one star.

Doublet. A pair of associated lines in a *spectrum* characteristic of the *alkali metals.*

Drachm Fluid. British unit of *volume* ; 60 *minims*; 3.55 cm^3.

Drain. The electrode in *field-effect transistor* through which *charge charriers* leave the inter-electrode region.

Drug. 1. Chemical substance used in medicine to cure or prevent disease. **2.** A habit-forming *narcotic* substance that causes physiological or emotional dependence.

Dry Cell. Small cell made of zinc metal containing paste of ammonium chloride zinc container acts as anode and a carbon rod as carthode; used in portable electronic appliances such as calculators, clocks torches, transistors etc.

Dry Distillation. Heating a solid to make it give vapour which is condensed into liquid.

Dry Ice. Solid carbon dioxide (CO_2), having temperature of —79° to—80°C; used for refrigerating food-stuffs.

Drying Oil. An animal or vegetable *oil* that harden, to a tough film when a thin layer is exposed to the air. The hardening is due to *oxidation* or *polymerization* of the *unsaturated fatty acids*. They are used in *paints* and varnishes (*e.g.*, *linseed oil*, dehydrated *castor oil*, and certain fish oils).

Ductility. A property of *metals* of being drawn out into a wire.

Dulong and Petit's Law. The product of the *relative atomic mass* and the *specific heat capacity*, *i.e.*, the *molar heat capacity* of a solid element is a constant and is approximately equal to 25 joules per *mole per kelvin*.

Duplet. A pair of *electrons* shared between two *atoms* forming a single covalent bond.

Duralumin. Alloy of aluminium containing aluminium (95%), copper (4%) and small amounts of magnesium, manganese, silicon etc. It is very light and hard.

Dutch Gold. Coper-Zinc alloy, can be used as substitute for gold.

Dutch Liquid. Ethylene dichloride ($C_2H_4Cl_2$), an anaesthetic discovered by Dutch chemists.

Dwarfism. Stunted growth, generally due to some genetic defect, involving alteration of proportions of various parts of body. In plants it is usually caused by gibberellin deficiency.

Dyad. Bivalent atom, radical or element.

Dye. Colour; stain.

Dynamic Allotropy. Allotropy in which allotropes are in dynamic equilibrium with each other.

Dynamic Equilibrium. If two opposing processes are going on at the same rate in a system, thus keeping the system unchanged, the system is said to be in dynamic equilibrium.

Dynamite. Powerful explosive, generally consisting of an absorbant saturated with nitro-glycerin.

Dynamo. Machine used to generate electric current by means of relative movements of conductors and magnets.

Dynamometer. Instrument used for measuring energy, force or power.

Dynatron. Refers to four-electrode vacuum tube used in oscillators.

Dynatron Oscillator. Oscillator using tetrode in such a way that anode current increases as anode voltage is reduced.

Dyne. Unit of force. It is the force that acting upon one gram of mass for one second produces velocity of one centimetre per second.

Dynode. Electrode whose primary functions is secondary emission of electrons.

E

Ear. Organ of hearing and balance in vertebrates. Hearing is a sensation produced by vibrations or sound waves which are converted into nerve impulses by the ear and transmitted to the brain.

Earshot. Distance at which a sound can be heard easily.

Earth. A *planet* having its *orbit* between those of *Venus* and *Mars*. It is a sphere, slightly flattened towards the poles. Mean *density* 5.52×10^3 kg m^{-3}; mass 5.976×10^{24} kg. The Earth consists of a gaseous *atmosphere*, a liquid *hydrosphere*, and a solid lithosphere.

Earthing. A conductor that makes an electrical connection between the *conductor* and the Earth; the Earth is assumed to have zero *potential.*

Earth Sciences. The sciences related to the study of the *Earth*, including *geology* (*geophysics* and *geochemistry*), *geography, meteorology* and *oceanography.*

Earthshine. A faint illumination of the dark side of the *Moon* during a crescent *phase*, due to sunlight reflected from the Earth's surface.

Ebonite. A hard black insulating material made by vulcanizing *rubber* with high proportions of sulphur. It contains about 30% combined sulphur.

Ebullimeter. It is an instrument which is used to measure accurately the boiling points of solutions.

Ecdysone. Steroid hormone secreted by prothoracic glands of insets for initiating ecdysis.

Echo. Effect produced when sound or other radiation is reflected on striking a solid obstacle like wall or rock.

Echo-chamber. Reverbernt room in studio used for producing and adding echo effects to sound that is being recorded.

Echolocation. The location of an object by determining the direction of an *echo* reflected from it, or the time taken for the echo to return *e.g,. radar, echo sounding etc.*

Echo sounding. A method of estimating the depth of the sea beneath a ship by measuring the time taken for a *sound* pulse to reach the sea bed and for its *echo* to return.

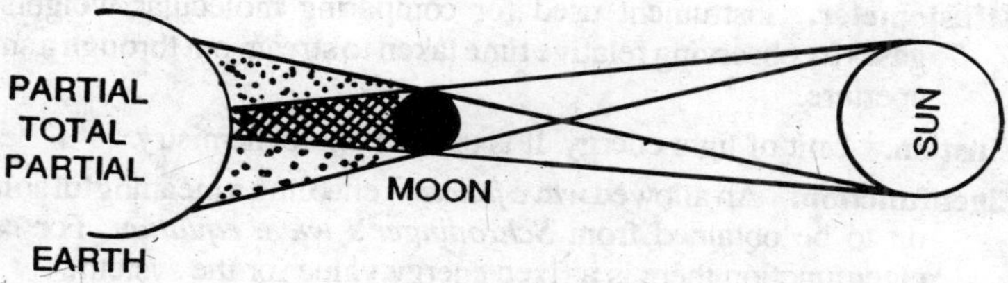

Fig. 7.

Eclipse. The passage of a non-luminous body into the *shadow* of another. An 'eclipse of the *Moon*', or lunar eclipse, occurs when the Sun, the Earth, and the Moon are in line so that the shadow of the Earth falls upon the Moon. An 'eclipse of the Sun', or solar eclipse, is said to occur when the shadow of the Moon falls on the Earth.

Ecology. Study of living organisms in relation to environment.

Ecosystem. Interacting system of living organisms in an area with their physical environment.

Eczema. Inflammatory skin disease characterised by watery discharge and formation of crust or scales.

Edentate. Having no teeth *e.g.*, ant-eaters, sloths.

Edison Accumulator. Name given to a nickel-iron accumulator.

E.D.T.A. Ethylene-diaminetetraacetic acid.

$(NCH_2CH_2N)_2(HCOOH_2)_4$. It is used as a chelating agent.

Eel-spear. Instrument with broad prongs used for catching eel (fish).

Effective Resistance. The *resistance* of a *conductor* of electricity to *alternating currents*; in addition to the *direct current* resistance it includes the effect of any losses caused by the current.

Effector. A specialised animal tissue or organ that performs a response to an environmental stimulus, e.g., muscles, endocrine glands.

Efferent. Outgoing.

Effervescence. Escape of gas bubbles from a liquid, on account of chemical action.

Efficiency of a Machine. The ratio of the output *energy* to the input energy. The efficiency of a machine can never be greater than unity.

Efflorescence (chem.). The property of some crystalline *salts* of losing a part of their *water of crystallization.*

Effusion of Gases. The passage of *gases* through small apertures under pressure.

Effusiometer. Instrument used for comparing molecular weights of gases by observing relative time taken to stream out through a small aperture.

Einstein. Unit of light energy. It is used in photochemistry.

Eigenfunction. An allowed *wave function* enabling a meaningful solution to be obtained from *Schrodinger's wave equation.* For each eigenfunction there is a fixed energy value for the system.

Einsteinium Es. *Transuranic element,* At. No. 99. The most stable *isotope*, einstemium-254, has a *half-life* of 276 days.

Einstein's equation. $E = mc^2$, where E is the energy equivalent of a mass m, and c is the *speed of light.* A direct consequence of Einstein's special theory of *relativity*, this equation is the basis of all *nuclear energy.*

Einstein Number. It is a dimensionless number. It is used in magneto fluid dynamics. It is equal to the ratio of velocity of a fluid to the speed of hight.

Einstein Shift. The slight shift towards the red of the lines of a solar spectrum. It is due to sun's gravitational field.

Elastance. The *reciprocal* of *capacitance*; it is measured in reciprocal *farads* or *darafs.*

Elastic Collision. A collision between bodies under ideal conditions, such that their total *kinetic energy* before collision equals their total kinetic energy after collision. In *nuclear physics* an elastic collision is one in which an incoming particle is scattered without causing the *excitation* or breaking up of the struck *nucleus.*

Electric. Pertaining to electricity.

Electric Arc. Luminous space between electrodes when a current passes across them.

Electric Chair. Chair on which condemned criminals are put to death using electric current.

Electric Constant. Permittivity of free space. εo. The fundamental constant that has the value 8.85416×10^{-12} farad per metre. It arises as the constant of proportionality in *Coulomb's law.*

Electric Current. A flow of *electric charge* through a *conductor.* The *charge carriers* may be *electrons, ions,* or *holes.* The *SI unit* of current is the *ampere.* Symbol *I.*

Electric Current, Heating Effect of. When an *electric current* flows through a *conductor*, the quantity of heat produced is proportional to the resistance of the conductor and is equal to *VI, V* being the *potential difference* in *volts, I* the current in *amperes,* and *R* the resistance in *ohms.*

Electric Displacement. Electric flux density. The *electric charge* per unit area displaced across a layer of conductor placed in an *electric field.*

Electric Double Layer. Diffused aggregation of positive and negative charges surrounding suspended colloidal particles that help in maintaining stability.

Electric Eye. Photo-electric cell.

Electric Field. The region surroundings an *electric charge,* in which a *force* is exerted on a charged particle. The electric field strength *E,* or *force* exerted upon a unit charge at a distance *r* from a charge *Q,* is given by:

$$E = Q/4\pi r^2\varepsilon$$

where ε is the *permittivity*. For free space (i.e., a vacuum) ε becomes ε_0, the *electric constant*, and has the value 8.854185×10^{-12} F m^{-1}. The strength of the field can also be described by the *electric displacement D*. In a vacuum $D/E = \varepsilon_0$.

Electric Flux. The quantity of electricity displaced across unit area of a dielectric. It is the scalar product of the *electric displacement* and the area.

Electricity. Power or energy produced by physical phenomenon involving electrons and protons; used for light fans and a wide range of electronic appliances.

Electric Probe. Devise used for measuring density of electrons and ions, space and wall potentials, electron temperatures and random electron currents in a plasma.

Electric Spark. Discharge of electricity through an insulator accompanied by light and sound.

Electro-Acoustics. Conversion of electric energy into sound energy and *vice-versa.*

Electro-analysis. Separation by using process of electrolysis.

Electro-biology. Science of electrical phenomenon in living organisms.

Electro-cardiogram (ECG). Photographic record of electrical variations that occur during contraction of heart muscles.

Electro-chemical Equivalent. Number of grams of element liberated by passage of one coulomb of electricity.

Electro-chemistry. Study of relationship between electricity and chemical changes.

Electroculture. Cultivation of plants under stimulus of electricity.

Electrocution. To kill by means of electricity.

Electrode. Conductor by which current of electricity enters or leaves electrolytic cell, gas discharge tube or thermionic valve. Positive electrode is called anode and negative electrode is cathode.

Electrodeposition. The process of depositing by *electrolysis*, the deposition of one *metal* on another as in *electroplating*.

Electrode Potential. The *electric potential* developed on an electrode that is in equilibrium with a solution of its *ions* Electrode potentials cannot be measured absolutely and are generally specified by comparison with a *hydrogen electrode*, which is assumed to have an electrode potential of zero.

Electrodialysis. A method of *desalination*. Water containing salt is fed into a *cell* containing two *electrodes* which are seperated by an array of *semipermeable membranes*, they are alternately semipermeable to positive and negative *ions*. The ions collect between alternate pairs of membranes enabling a stream of pure water to be extracted from the cell.

Electrodynamics. Study of relationship between electric and magnetic forces and their mechanical causes and effects.

Electrodynamometer. Instrument used for measuring current, voltage or power in direct current (DC) and alternate current (AC) circuits.

Electro-extraction. Recovery of metal from its salts by passing electric current through its solution.

Electro-kinetics. Branch of science that treats distribution of electric currents.

Electrolysis. Decomposition by electric current.

Electrolyte. A *compound* that, in *solution* or in the molten state, conducts an *electric current* and is simultaneously decomposed by it.

Electrolytic Capacitor (Condenser). A fixed electrical *capacitor* in which one *electrode* is a *metal* (usually aluminium) foil coated with a thin layer of the metal *oxide*, and the other electrode is a non-corrosive *salt solution* or paste. The metal foil is maintained positive to prevent the removal of the oxide film by the hydrogen liberated. The advantage of electrolytic capacitors is that they provide a high *capacitance* in a limited space.

Electrolytic Cell. A cell in which *electrolysis* can takes place on passing an *electric current* through an electrolyte.

Electrolytic Dissociation. The *dissociation* of the *molecules* of *electrolytes* into changed *ions*. The degree of dissociation determines the electrical *conductivity* of the solution and also other properties.

Electrolytics Gas. Detonating gas. A mixture of hydrogen and oxygen, in a ratio of 2 to 1 by *volume.*

Electrolytic Rectifier. A *rectifier* consisting of two *electrodes* immersed in an *electrolyte*. It is used to convert an *alternating current* into a *direct current*. It depends on the properties of certain *metal* and *solutions* to allow current to flow in one direction only.

Electrolytic Refining. The purification of copper and some other metals by *electrolysis.*

Electromagnet. Coil of wire having soft iron core, when electric current flows through coil, iron core behaves like a magnet but as soon as current stops flowing it loses its magnetism.

Electromagnetic Environment. Radio frequency fields existing in a given area or space.

Electromagnetic Spectrum. Total range of wavelengths or frequencies of electromagnetic radiation.

Electromagnetic Theory. Maxwell's doctrine explaining light in terms of electromagnetic waves.

Electromagnetic Wave. Disturbance which propagates outward from any electric charge which oscillates or is accelerated.

Electro-magnetism. Branch of science that deals with relation of electricity to megnetism.

Electromechanical. Mechanical process or device making use of electricity.

Electrometallurgy. Electric processes used in separating a metal from its ore.

Electromotor. Apparatus for applying electricity as a motive power.

Electron. An *elementary particle* classed as a *lepton* having a *rest mass* of 9.109558×10^{-31} *kilogram*, approximately 1/1836 of that of a hydrogen *atom*, and bearing a negative *electric charge* of $1.602\,192 \times 10^{-19}$ *coulomb.* The radius of the electron is $2.817\,77 \times 10^{-15}$ *metre*. The positively charged *anti-particle* of the electron is the *positron*, and the word 'electron' is sometimes used to include both negative electrons (negatrons or negatons) and positive electrons (positrons or positons).

Electron Affinity. 1. The tendency of an *atom* or *molecule* to accept an *electron* and form a negative *ion*. **2.** The *energy* liberated when one *mole* of any gaseous *element* is converted into negative ions. The *halogens* have high electron affinites.

Electron Capture. 1. The formation of a negative *ion* when a *free electron* is captured by an *atom* or *molecule* **2.** A *radioactive transformation* as a result of which a *nucleus* captures an inner *orbital electron* and the resulting excited ion emits an *X-ray photon.*

Electron Density. The density of *electronic charge* at a given point in a *molecule* also defined as the probability of finding an *electron* at the particular point.

Electron Diffraction. A *diffraction* effect resulting from the passage of *electrons* through *matter*. The phenomenon of electron diffraction is the principal evidence for the existence of waves associated with electrons. The diffraction of electrons when passed through *crystals* or thin *metal* foils is used as a method of investigating crystal structure.

Electronegative Elements and Groups. Elements and groups that take up *electrons*, thus acquiring a negative *electric charge*, when united with other radicals by electrovalent bonds. The *halogens,* oxygen, sulphur, and other *non-metallic elements* are generally electronegative.

Electron Gun. The source of *electrons* in a *cathode-ray tube* or *electron microscope*. It consists of a *cathode* emitter of electrons, an *anode* with an aperture through which the beam of electrons can pass, and one or more focusing and control *electrode.*

Electronics. Physical science concerned with the development of electrical circuits using *semiconductors, thermionic valves,* and other devices in which the motion of *electrons* is controlled for purposes of communications, control, or computing.

Electron Lens. A system of electric or *magnetic fields* used to focus a beam of *electrons*. They are used in *electron microscopes, cathode-ray tubes,* etc.

Electron Micrograph. A photograph of an object obtained with an *electron microscope.*

Electron Microscope. An instrument similar to the ordinary *light microscope*, but having a much greater *resolving power.* Instead of a beam of light to illuminate the object, a parallel beam of *electrons* from an *electron gun* is used.

Electron Multiplier. *Photomultiplier.*

Electron Optics. Refers to the study of the methods used to focus beams of *electrons* by means of *electron lenses*, etc.

Electron-probe Microanalysis. A method of analysing a very small quantity of a substance by directing a finely focused *electron* beam on to it so that an *X- ray* emission is produced characteristic of the *elements* present in the sample. The method may be used quantitatively for elements whose *atomic numbers* exceed 11.

Electron-spin Resonance ESR. A phenomenon which is exhibited by paramagnetic substance due to their unpaired *electrons*. The spin of an unpaired electron is associated with a *magnetic moment* which may align itself in one of two ways with respect to an applied *magnetic field*, each possible alignment corresponding to a different *energy level*. By applying magnetic field and irradiating the specimen with suitable *electromagnetic radiation*, transitions between these two energy levels can be made, falling in the *microwave* region of the *electromagnetic spectrum*, thereby producing the phenomenon known as electron-spin resonance. Electron-spin resonance spectroscopy consists of analysing this *hyperfine structure* so that the electron can be located within the molecule, thus providing information about the molecule's structure. ESR is also used for studying *free radicals*.

Electron-volt. eV. A unit of energy widely used in *nuclear physics*. The increase in energy or the *work* done on an *electron* when passing through a *potential* rise of 1 *volt*. 1 electron-volt $=1.602 \times 10^{-19}$ *joule*. I MeV = 10^6 electron-volts ; 1 GeV = 10^9 electron-volts.

Electrophilic Reagents. Cationoid reagents. *Reagents* that react at centres of high *electron density*. They are essentially electron acceptors (*e.g.*, *halogens*) that gain or share electrons from an outside *atom* or *ion*.

Electrophoresis. Cataphoresis. The migration of the electrically charged *solute* particles present in a *colloidal solution* towards the oppositely charged *electrode*, when two electrodes are placed in the solution and connected externally to a source of *E.M.F.*

Electrophorus. A laboratory apparatus for showing electrostatic charging by *induction*.

Electroplating. Depositing a layer of *metal* by *electrolysis*, the object to be plated forming the *cathode* in an electrolytic tank or both containing a *solution* of a *salt* of the metal that is to be deposited.

Electropositive Elements and Groups. Elements and groups which give up *electrons*, thus acquiring a positive *electric charge*, when united with other radicals by electrovalent bonds. The *metal* and *acidic hydrogen* are generally electropositive.

Electroscope. An instrument for detecting the presence of an electric *charge.*

Electrostatic Field. Refer to a region in which a stationary electrically charged particle would be subjected to a *force* of attraction as a result of the presence of another stationary *electric charge.*

Electrostatic Generator. A machine designed for the continuous separation of *electric charge*. Examples include the *Wimshurst machine* and the *Van de Graaff generator.*

Electrostatic Precipitation. A widely used method of controlling the pollution of air. The gas, containing *solid* or *liquid* particles suspended in it, is subjected to a uni-directional *electrostatic field*, so that the particles are attracted to, and deposited upon, the positive *electrode.*

Electrostatics. The study of static *electric charge* and the *forces* and *fields* associated with them.

Electrostatic Units. ESU. A system of electrical *units* based upon the electrostatic units of *electric charge.*

Electrostriction. The change in the dimensions of a *dielectric* when placed in an *electric field.*

Electrotyping. Refers to the production of copies of plates of type, etc., by the electrolytic deposition of a layer of *metal* on a previously prepared mould. This is a cast of the object to the copied, made of *plastic* material and coated with a layer of *graphite*, which acts as a *conductors* of electricity. It is then suspended to act as a cathode in an electrolytic bath containing a *solution* of a salt of the metal required, generally, copper.

Elevation of Boiling Point. The rise in the *boiling point* of a *solution* which is produced by a non-volatile substance dissolved in a *solvent*. For a *dilute* solution the elevation is proportional to the number of *molecules* or *ions* present and the elevation produced by the same *molecular concentration* is a constant for a particular solvent. This forms the principle of the boiling point method for the determination of *relative molecular masses.*

Eleven-year Period. A periodic change in the occurrence of *sunspots*, the cycle being complete in approximately eleven years; associated with this is a cycle variation in the *daily variation.*

Elinvar. A variety of *steel* containing 86% nickel and 12% chromium. The *elasticity* is almost unchanged by change of *temperature*; it is used for hairsprings of watches.

Elliposid. A solid figure traced out by an *ellipse* rotating about one of its *axes*. A prolate ellipsoid is produced if it is rotated about the major axis and an oblate elliposoid about the minor axis.

Elliptically Polarized Light. *Light* that can be resolved into two vibrations lying in *planes* at right angles, and of equal *frequency*. The form of this ellipse is determined by the *amplitude* of these two vibrations and by the difference of *phase* between them.

Elution. The removal of an *adsorbate* from and adsorbent by dissolving it in a *liquid* known as eluent. The resulting solution is called the eluate.

Elutriation. The washing, separation, or sizing of fine particles of different mass by suspending them in a current of air or *water.*

Emanation. Radium emanation. The *gas* formed by the *radioactive disintegration* of certain substances, consisting principally of *radon, thoron* and *action.*

Emery. A mixture of *corundum* and iron oxide, usually *magnetite* Fe_3O_4 used as an *abrasive.*

Emetine. $C_{29}H_{40}O_4N_2$. An alkaloid obtained from the roots of Brazilian ipecacuanha, m.p. 68°C., used as an emetic and a remedy for amoebic dysentery.

EMF. *Electromotive force.*

Emission of Radiation. The rate at which a body emits *heat radiation* to its surroundings. It depends on the *temperature* of the body, the temperature of its surroundings, and the nature of the surface of the body.

Emission Spectrum. The *spectrum* observed when electromagnetic radiation coming from a source is examined with a *spectroscope.*

Emigration. Movement of some individual out of a population.

Emitter. It is one of the three *electrodes* in a *transistor.*

Empirical. Based upon the results of experiment and observation only.

Empirical Formula. The simplest type of chemical *formula*, giving the proportion of each *element* present.

Emulsifier. Emulsifying agent. A substance, small quantities of which help to form or stabilize an *emulsion*. *Detergents* are frequently used as emulsifiers.

Energy Flex. It is the rate of flow of energy per unit area.

Energy Levels. An *atom, molecule*, or *nucleus* can exist only in certain definite states (*quantum states*) each having definitive energies. Thus, for each different atom or nucleus, there exists a series of energy levels corresponding to these permissible states. The lowest stable energy level is referred to as the *ground state*; system at higher energy levels than the ground state are said to be excited.

Energy-rich Bonds. A term used in *biochemistry* to distinguish between chemical *bonds* that when broken yield a large amount of *free energy* and those that give only a small yield of free energy (energy-poor bonds'). The energy referred to in this context is the free energy liberated on *hydrolysis.*

Energy Value of a Food. A measure of the energy made available by the complete combustion of a stated *mass* of the food; bond between carbon atoms in its molecules, it is often given in joules per kilogram or large *Calories* per pound.

Energy. A device for converting one form of *energy* into another.

Enols Organic compounds containing the group — CH=C(OH) —.

Enrich. To increase the *abundance* of a particular *isotope* in a mixture of isotopes, particular to increase the abundance of the *fissile* isotope of a *nuclear fuel.*

Environment. Physical, chemical and biological factors taken together to which an organism is exposed.

Enyne. When a hydrocarbon contains a double and a triple bond between carbon atoms in its molecule, it is known as enyne.

Enzyme. Refers to organic substances that brings about a chemical change and itself does not undergo any change.

Enzymolysis. The *decomposition* of a substance catalyzed by *an enzyme.*

Eosin. Tetrabromofeuorescin. $C_{20}H_8Br_4O_5$. A red crystalline *insoluble* substance obtained from *fluorescein*, m.p. 295-6°C., used as a red *dye.*

Epact. 1. The difference in days between the length of solar year and a lunar *year*. **2**. The *Moon's* age in days at the start of the calender year.

Ephedrine. $C_6H_5CHOHCH(CH_3)NHCH_3$. A white crystalline *optically active alkaloid*, m.p. 40°C., used in medicine to treat asthma, colds, etc.

Ephemeris. A table showing the predicted positions and the movements of celestial body such as a *planet* or *comet*. An annual publication containing astronomical data.

Emulsion. A *colloidal solution* in which the *dispersed phase* consists of minutes droplets of *liquid.*

Emulsion, Photographic. The light-sensitive coating on a *film* or plate. A 'nuclear emulsion' is a photographic emulsion specially prepared to record the tracks of *elementary particles* and nuclear fragments that pass through it.

Encephalograph. Refers to X-ray photograph of brain.

Encyclopaedia. Book(s) containing alphabetically arranged comprehensive information covering entire circle of knowledge and leanring.

Endemic. That which is prevalent or regularly occurring in particular area.

Endergonic. Process characterised by concommitant adsorption of energy *e.g.*, protein synthesis.

Endo-. Meaning inside, within.

Endocrine glands. Ductless glands; number of such glands are present in the body such as, adrenals, pituitary, thyroid etc.

Endoenzyme. An *enzyme* that remains within a living *cell* and does not diffuse through the cell wall.

Endoergic Process. An *endothermic process* (used in the context of a *nuclear reaction*).

Endosmosis. Movement of solvent through semipermeable membrane from exterior to interior of cell.

Endosome. Central mass consisting of mainly chromatin material found in nucleus of certain protozoa.

Endosulfan ($C_9H_6Cl_6O_2S$). Brown crystalline solid, insoluble in water but soluble in xylene; used as insecticide, acts as contact and stomach poison.

Endothermic Reaction. Chemical reaction requiring heat/energy/ protein synthesis.

Endrin ($C_6H_8Cl_6O$). One of the most toxic hydrocarbon insecticide, insoluble in water, slightly soluble in alcohols and petroleum; due to high residual toxicity, its use as insecticide has been discouraged.

Energy. The capacity for doing *work*. Interconversion between various forms of energy can only occur in the presence of *matter*. Energy can only exist in the absence of matter in the form of radiant energy. The derived *SI unit* of energy is the *joule*. Symbol *E*.

Energy Bands. *Orbital electrons* are associated with specific amounts of energy, the change from one *energy level* to another taking place in *quantized steps*. In a crystalline *solid* the energies of all the electrons and *atoms* fall into several 'allowed' energy bands which lie 'forbidden' bands.

Epicentre. The point on the surface of the Earth that lies directly above the focus of an earthquake.

Epichlorohydrin. 1-Chloro-2,3-epoxypropane. $\overline{OCH_2CH}CH_2Cl$. A colourless liquid, b.p. 116°C. A highly reactive *epoxy* compound used in the manufacture of *epoxy resins.*

Epimerism. A type of *optical isomerism* occurs in *carbohydrates* and some other types of compound. It is because of the formation of isomers that differ in their molecular arrangements about an asymmetric atom in a molecule containing two or more asymmetric atoms.

Epoxide. A *compound* having an oxygen *atom* bound to two carbon atoms, forming a three-membered ring.

Epoxy-. Prefix denoting an oxygen atom whose free *valences* are attached to different atoms, which are otherwise connected: applied to cases in which oxygen forms a three membered ring with two carbon atoms.

Epoxyethane. Ethylene oxide, oxirane. $\overline{CH_2CH_2O}$. A colourless flammable gas made by the *oxidation* of *ethene* in the presence of a *catalyst.* It is used to make *ethanediol, ethanolamines*, etc.

Epoxy Resins. *Thermosetting resins* obtained by the reaction of *epichlorohydrin* with *polyphydric* compounds such as bisphenol A (4, 4′- isopropylidenediphenol), in the presence of a *catalyst.* These are used in the manufacture of electrical components, as structural materials, in surface coatings, and as *adhesives.*

Epsom Sales. Eposomite. *Magnesium sulphate.* $MgSO_4 . 7\ H_2O$.

Equation, Chemical. A representation of a *chemical reaction,* with the help of *symbols* of the *elements* to represent the actual *atoms* and *molecules* taking part in the reaction. *e.g.*, the chemical equation.

$$H_2 + Cl_2 = 2HCl$$

represents the reaction between hydrogen and chlorine to form *hydrogen chloride.*

Equation, Mathematical. A statement of equality between known and unknown quantities. Thus, the equation $3x = 15$ is true only when $x = 5$.

Equation of State. A relation between the *pressure, p, volume V*, and *thermodynamic temperature T* of a substance. *Van der Waals equation of state.*

Equation of Time. The difference between mean solar time, as given by a clock, and apparent solar time, *i.e.*, sundial time.

Equator, Terrestrial. The *great circle* of the Earth, lying in a *plane* perpendicular to the *axis* of the Earth, that is equidistant from the two Poles.

Equilateral Figure. A figure having all its sides equal in length *e.g.*, an equilateral *triangle.*

Equilibrium. A state of balance between opposing *forces* or effects.

Equilibrium Constant. In any *chemical reaction* there is always a state of *chemical equilibrium*, at a given *temperature* and *pressure,* between the *concentration* of the reactants and the concentration of the products. The position of this equilibrium constant, *K*, such that in the reaction:

$$aA + bB = cC + dD$$

K is given by :

$$K_c = (C_C)^c (C_D)^d / (C_A)^a / (C_B)^b.$$

where C_A is the concentration of the substance A, '*a*' *molecules* of which take part in the reaction. In gas-phase reactions, *partial pressures* can be used instead of concentrations.

Equimolecular Mixture. A *mixture* containing substances in equal molecular proportions; *i.e.*, in the ratio of their *relative molecular masses.*

Equivalent. Quantity of substance that is equal to combining value of an arbitrarily fixed quantity of another substance in particular reaction.

Ethanal (CH_3CHO). Colourless, pungent smelling liquid miscible with ethanol, either and water; acetaldehyde.

Ethane (C_2H_6). Hydrocarbon of methane series.

Ethanol (C_2H_5OH). Ethyl-alcohol, colourless liquid, miscible with water obtained by fermentation of molasses or starch; used in preparation of ether, ester, chloral, chloroform etc., and as solvent for paints, resins, varnishes etc.

Ether ($C_2H_5O.C_2H_5$). Clear upper air above clouds. Refers to a compound which is a fragrant smelling, colourless transparent volatile liquid of great mobility and high refractive power; can be prepared by dehydrating ethanol by means of concentrated sulphuric acid; used as anaesthetic and as solvent.

Ethers. Group of organic compounds (R–O–R′) produced by condensation of two alcohol molecules.

Ethion ($C_9H_{22}O_4P_2S_4$). Organophosphate which is slightly soluble in water, miscible in xylene, kerosene oil etc., used as acaricide and insecticide.

Ethnology. Science relating to races of mankind and thier relation to one another.

Ethyl (C_2H_5). Alkyl radical of ethane.

Ethylamine ($NH_2C_2H_5$). Ammonia like substance with one atom of hydrogen replaced by ethyl.

Ethylene (C_2H_4). Colourless, gas, sparingly soluble in water burns with smoky flame, boiling point 105°C used for ripening fruits artificially.

Ethyl Ethanoate. Ethyl acetate. $CH_3COOC_2H_5$. A colourless liquid with a pleasant fruity smell, b.p. 77°C., it is used as a *solvent* and in flavourings and perfume.

Ethyl Fluid. A *Solution of tetraethyllead*, $Pb(C_2H_5)_4$, and *dibromoethane* $C_2H_5Br_2$, used as an anti-*knock* compound in motor *fuel.*

Ethyl Group. *Univalent alkyl radical* C_2H_5.

Ethyl Nitrite. Nitrous ether $C_2H_5NO_2$. A *volatile liquid* with a sweet smell b.p. 17°C, used in medicine.

Ethyne. Acetylene. A colourless poisonous *gas*; the first member of the *alkyne* series. It is made by the action of water on *calcium dicarbide.* It is used as a staring material for many *organic compounds,* and for *welding.*

Etiolate. To grow pale with long whitish internodes and small yellow leaves, for want of proper light.

Eucarpic. Having separate sexual and asexual reproductive organs, functioning.

Euchlorine. Mixture of cholorine and chlorine Peroxide, green coloured highly explosive gas.

Eudiometer. A *glass* tube used for measuring volume change in *chemical reactions* between gases.

Eugenics. The study of the genetic control of human populations, with a view to improving their constitution, by selectively encourging breeding among those people considered by eugenicists to be the most desirable.

Eugenol. ($C_{10}H_{12}O_2$) A colourless oily *liquid,* m.p. 9.2°C., b.p. 255°C., can be extracted from oil of cloves. It is used in perfumes and as an *antiseptic.*

Europium. Eu. Element R.a.m. 151.96. At No. 63. A silvery metal, r.d. 5.24, m.p. 822°C., b.p. 1529°C.

Eutectic Mixture. A *solid solution* of two or more substances, having the lowest *freezing point* of all the possible mixture of the components.

Eutectic Point. The minimum freezing point attainable, corresponding to the *eutetic mixture*, is the eutectic point.

Evaporation. The conversion of a *liquid* into *vapour*, without necessarily reaching the *boiling point*, it is used in concentrating *solutions* by evaporating off the *solvent*.

Even-even Nucleus. A nucleus containing both an even number of *protons* and an even number of *neutrons*.

Even-odd Nucleus. A nucleus containing an even number of *protons* but an odd number of *neutrons*.

Excess Electron. Electron in a semiconductor donated by an impurity, which is not required in bonding system of crystal and is thus available for conduction.

Excimer. An exited dimer, formed by association of excited and unexcited molecules which in ground state would remain dissociated.

Excitation. Addition of energy to atom, molecule or nucleus, transferring it from its ground state to higher energy level.

Exciton. Non-conduction, non-localised, excited electron state in a semiconductor.

Exergonic. Chemical reaction yielding energy.

Exoenzyme. Enzyme that functions outside the cell in which it is produced *e.g.*, pepsin.

Exon. Polynucleotide sequence that can code for specific protein.

Exosmosis. Passage of solvent molecules through semipermeable membrance from interior of cell to exterior of cell.

Exothermic Process. Process in which heat energy is released.

Exotic. Outlandish; not native, introduced from foreign country.

Exotoxin. Toxic substances produced in cells of many Gram-positive bacteria which are instantly released through bacterial wall into the surroundings.

Explosion. Violent and sudden expansion of gases produced by rapid combustion that exerts a very strong force when enclosed in a small space.

Extensometer. Instrument used for measuring extension produced in a body under an applied stress.

Extraordinary Ray. A ray of light that is bent on entering double-refracting crystal.

Eye. An organ of sight or vision.

Eye-piece. Magnifying lens of system of lenses in optical instrument, used to view the image forming lens (es).

F

F1 Generation. The first generation of progeny obtained in breeding experiments. Successive generations are called F_2, etc.

Face-lifting. Operation that involves smoothing and firming the face.

Factor, Prime. The factor of a quantity are the ***prime numbers*** that, when multiplied together, give the quantity. Thus, the prime factors of 165 are 3, 5 and 11.

Factorial. The *product* of a number and all the consecutive positive whole numbers below it down to 1. Thus, factorial 6, written ∠6 or

$$6 \times 5 \times 4 \times 3 \times 2 \times 1 = 720.$$

Faculae. Large bright areas of the *photosphere* of the *Sun*, whose *temperatures* are higher than the average of the Sun's surface.

Fad (flavin adenine dinucleotide). Coenzyme formed by the condensation of riboflavin phosphate adenylic acid.

Falciform. Sickle-shaped, leaves.

Family. Unit used in the classification of living organisms consisting of one or more genera (singular genus).

Faraday. Quantity of electricity that liberates or deposits one gram-equivalent of ion.

Fatigue. Build up of waste products, ability of muscles to contract being obliterated or greatly reduced.

Fats. Organic compounds containing carbon (C), hydrogen (H) and oxygen(O); esters of fatty acids and alcohol.

Feedback. In general, the coupling of the output of a process to the input.

Fehling's Solution. A solution of *copper* (II) *sulphate*, $SuSO_4$, *sodium hydroxide*, NaOH, and potassium sodium tartrate (*I ochelle solt*). Used for the detection and estimation of certain *sugars, aldehydes,* and other *reducing agents* which produce a red precipitate of *copper* (I) *oxide* Cu_2O.

Feldspar. A large group of rock-forming *minerals*. It consists chiefly of *alumino-silicates* of potassium and sodium. They are constituents of *granite* and other primary rocks.

Fensulfothin ($C_{11}H_{17}O_4PS_2$). Yellowish-brown liquid, boiling point 140°C, soluble in most organic solvents, used as insecticide and nematicide, system and *i.e*,. contact position prolonged residual effect.

Fenthion ($C_{10}H_{15}O_3PS_2$). Colourless oily liquid, boiling point 87°C., not soluble in water but readily dissolves in most organic solvents; used in insecticide, contact and stomach poison, has long residual effect.

Fermentation. Partial breakdown, in anaerobic conditions of glucose or other organic substances by living organisms to yield ethanol and carbon dioxide.

Ferritin. Large protein molecule opaque to electrons. It is used as marker in electron microscope; normally present in spleen as storage protein for iron.

Fertilizer. Substance added to soil to increase the quantity or quality of plant growth. Fertilizers replenish the soil and are of two types:

(i) Organic fertilizers such as sewage.

(ii) Inorganic fertilizers such as ammonium sulphate.

FFA (free fatty acids). Extremely labile circulating fat components. They have half life period of only few minutes.

Fibrin. Insoluble protein precipitated as network of fibres on blood coagulation in vertebrates.

Filtrate. Clear liquid obtained after filtration, it has no suspended matter.

Filtration. Process used for separating insoluble solid particles from liquid by passing the fluid through porous material (filter paper etc.).

Fire-damp. Explosive mixture of methane (CH_4) and air, formed in coal mines.

Fire-extinguisher. Apparatus used to put out fire.

Fire-fly. Insect (beetle) that emits light at night; luminous organs are situated on VI and VII abdominal segments wherein luciferin is oxidised with help of enzyme luciferase producing cold-light or chemoluminiscence.

Fittig Reaction. Wurtz-Fittig synthesis. The synthesis of *alkylarene* hydrocarbons by the action of metallic sodium with a mixture of a *haloalkane* and a halogenated *benzene* derivative.

Eitzgerald-lorentz Contraction. The hypothesis put forward to explain the result of the *Michelson Morley experiment on* the supposition that a body moving with high *velocity* through the *ether* would experience a contraction in length in the direction of the motion.

Fixation of Atmospheric Nitrogen. The manufacture of useful *compounds* of nitrogen from the atmospheric nitrogen. The first practical process was the *Birkeland and Eyde process*; the *Haber* and *Serpek processes* are now the main ones used. In addition, certain *bacteria* in the soil fix atmospheric nitrogen.

Fixed Air. (former name) *carbon dioxide*, CO_2.

Fixed Alkali. Used for *potassium* or *sodium carbonate*, to distinguish them from volatile alkali, ammonium carbonate.

Fixed Oil. Non-volatile oil found in plants. It is used for cooking purposes.

Fixed Stars. True *stars*: heavenly bodies that do not appear to alter their relative positions on the *celestial sphere* compared to the planets, formerly known as 'wandering stars'.

Fixing, Photographic. Rendering that portion of the sensitive *film*, plate, or paper that has not been affected by *light*, insensitive to exposure, after *developing*. It is usually done by the action of *sodium thiousulphate*, $Na_2S_2O_3$ (*hypo*), which reacts with the unaffacted *silver bromide* to give a *soluble*, silver sodium thiosulphate, which is then washed away with water.

Flame. The glowing mass of *gas* produced during *combustion.*

Flame Photometry. A development of the *flame test* used in qualitative analysis; photometric measurement of flame *emission* is used to determine the concentration of substances introduced into the flame.

Flametest. A *qualitative* test for the detection of an element by the colour it or its compounds impart to a *Bunsen burner* flame. Sodium compounds give a bright yellow; potassium, caesium, and rubidium give a violet colour; strontium and lithium a red colour; barium, copper, thallium, and tellurium give a green colour.

Flash Point. Lowest temperature at which sufficient vapours are given by flammable liquid to ignite in presence of spark.

Flint. Impure silica (SiO_2). Flints used in lighters are made of alloys of metals like cerium and iron.

Flotation, principle of. The *mass* of *liquid* displaced by a floating body is equal to the mass of the body.

Flotation Process. The separation of a *mixture, e.g.,* of *zinc blende*, ZnS, and *galena*, PbS, based on the *surface tension* of *water.* Zinc blende is not easily wetted by water and flats, supported by the surface film of water, while galena sink.

Flue Gas. Gaseous product of combustion from boiler furnace, it contains carbon dioxide, carbon monoxide, oxygen, nitrogen and steam.

Fluidisation. In this technique, used in industrial chemistry, *(a)* mass of solid particles is brought into state of suspension by upward stream of gas blown through it in a reactor.

Fluorescien ($C_{20}H_{12}O_5$). Red crystalline organic compound dissolves in alkaline solutions producing a liquid of intense green fluorescence; used in dyes as an indicator.

Fluorescence. Refers to property of substance to emit electromagnetic radiations by absorbing energy.

Fluorine (F). Pale greenish-yellow gas, highly toxic; prepared by electrolysis of solution of potassium hydrogen fluoride in an hydrous hydrogen fluoride; used in pesticides.

Fluorocarbon. Compound derived from hydrocarbon by substitution of hydrogen atom with fluorine.

Fluoroscope. A florescent screen for the direct visual observation of *X ray* images; it is used in medicine.

Fluorosilicic acid. Hydrofluosilicic acid H_2SiF_6. An *acid* that is only stable in the form of its fuming *aqueous solution.* Used as a *disinfectant* and wood preservative.

Fluorspar. Calcium fluoride, CaF_2, colourless *crystals*, often coloured due to impurities. Used as a source of fluorine and its *compounds.*

Flux (chem.). A substance added to assist fusion.

Flux (phys.). The rate of flow of mass or energy per unit area normal to the direction of the flow. In *nuclear physics*, it is the product of the number of particles per unit volume and their average velocity.

Flux Density. The *magnetic flux* or *luminous flux* per unit of cross-sectional area. The *SI unit* of magnetic flux density is the tesla.

Fluxmeter. An instrument used for the measurement of *magnetic flux.* Essentially a moving coil *galvanometer* so designed that the coil experiences negligible restoring *torque* from its suspension system.

Fly Wheel. Large wheel having a heavy rim; it helps in maintaining constant speed in engines and machines.

F-number. Ratio of focal length to diameter of lens; smaller the F-number of lens, larger will be its aperture.

Foam. It is a *colloidal* suspension of a *gas* in a *liquid.* A solid foam (*e.g.*, foam rubber) is made by making a liquid foam and allowing it to solidify.

Focal Length. The distance from the *optical centre* or pole of a lens to the principal *focus.*

Focus. **1.** Focal point, (phys.) The point at which converging rays, of *light*, meet (real focus); or a point from which diverging rays are considered to be directed (virtual focus). **2.** (math). One of the fixed points used to define a curve, by a linear relationship with the distance from one of these fixed points to any point on the curve.

Folic Acid. One of water soluble vitamin found in green leafy vegetables its. Deficiency causes anaemia, sprue etc.

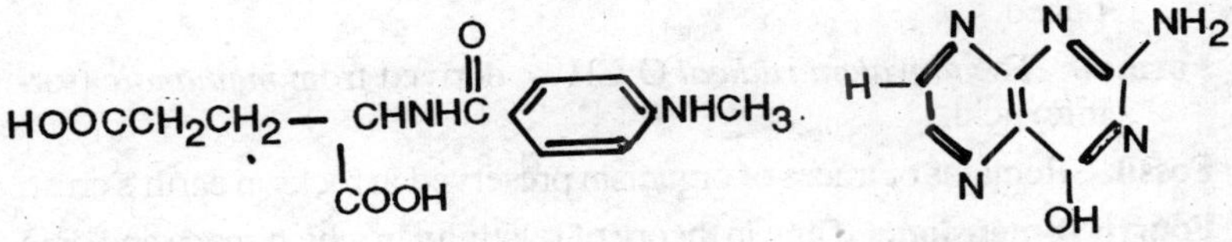

Folic acid

Food Preservation. The prevention of chemical *decomposition* and of the development of harmful *bacteria* in foods. It is generally effected by the sterilization of the food by heating in sealed vessels, *i.e.*, canning; or by making the conditions unfavourable for the development of bacteria, by pickling, drying, freezing, smoking, etc.

Food Value. Energy released on combustion of food.

Foot. British unit of length; one-third of a *yard*; 0.3048 *metre.*

Foot-candle. A unit of *illumination.* One *lumen* per square foot.

Foot-lambert. Unit of *luminance.* The luminance of a uniform diffuser emitting one *lumen* per square foot.

Foot-pound. A practical unit of *work.* The work done by a *force* of 1 pound weight acting through a distance of 1 foot.

Foot-poundal. A unit of *work* in the *f.p.s. system.* The work done by a *force* of 1 *poundal* acting through a distance of 1 foot.

Force F. An external agency capable of altering the state of rest or of uniform motion in a body. The force, F, required to produce an *acceleration*, a, in a *mass*, m, is given by $F=ma$. If m is in *kilograms*, a in m s^{-2}, F will be in *newtons.*

Formaldehyde. *Methanal.*

Formalin. A 40% *solution* of *methanal* containing *methanol* which acts as a stabilizer. It is used to preserve biological specimens and as a *disinfectant.*

Formic acid. *Methanoic acid.*

Formula (chem). The representation of a *molecule* or smallest portion of a *compound*, using *symbols* for the *atoms* of the *elements* which go to make up the molecule. *e.g.*, the formula of *water*, H_2O. The *empirical formula* of a compound is its simplest formula, indicating only the numerical ratio of the atoms present in a molecule, but not necessarily their actual number. Thus, the empirical formula of *hydrogen peroxide* is HO while its *molecular formula* is H_2O_2.

Formula (math. and phys.). A statement of facts in a symbolic form; by substitution a result applicable to particular data may be obtained.

Formyl. The *univalent radical* O:CH—, derived from *methanoic* (formic) acid.

Fossil. Remains or traces of organism preserved in rocks in earth's crust.

Fourth Dimension. Time in theory of relativity in which space and time are conceived as particular aspects of four-dimensional world.

Fraction. Mixture of liquids having points in a particular range.

Fractional Crystallization. The separation of a *mixture* of dissolved substances using their different *solubilities*. The *solution* containing the mixture is evaporated when the least *soluble* component crystallizes out first.

Fractional Distillation. Fractionation. The separation of a *mixture* of several *liquids* having different *boiling points*, 'fractions' boiling at different *temperatures* are collected separately.

Fractional Column. A long vertical column, containing rings, plates, of bubble caps, attached to a *still.* Due to internal *reflux* a gradual separation takes place between high and low boiling 'fractions' of a liquid *mixture.*

Fractionation. The separation of a *mixture*, of similar components, into fractions usually by *fractional distillation.*

Free Electron. Electron which is not attached to any atom, ion or molecule but is free to move under the influence of electric field.

Free Energy. Energy that is available for work in a system at a given temperature and pressure.

Free Radical. Atom or group of atoms having a single unpaired electron.

Freezing Point. Temperature at which a liquid is in equilibrium with a solid phase at standard pressure and below which liquid freezes or solidifies-freezing point of water is 0°C.

Frequency. Number of cycles, oscillations or vibrations of a wave motion in unit time (second).

Fresnel. Unit of frequency (= 10^{12} hertz).

Friable. Easily crumbled.

Friction. The *force* that offers resistance to relative motion between surfaces in contact. See *friction, coefficients of; rolling friction.*

Friction, Coefficients of. If F_3=the frictional resistance if a body is on the point of sliding along a specified surface, F_k=the frictional resistance when steady sliding has been attained, and R=the perpendicular *force* between the surfaces in contact, the static coefficient of friction =F_s/R; the kinetic coefficient = F_k/R.

Friedel -Crafts Reaction. Originally the *synthesis* of *aromatic hydrocarbons* by reacting *alkyl halides* with *benzene derivatives* in the presence of *anhydrous* aluminium chloride as a *catalyst.* It is now extended to include the addition of *alkenes* to, and the *condensation* of *alcohols* with, aromatic hydrocarbons in the presence of such catalysts as *anhydrous iron (III) chloride*, gallium chloride, boron trifluoride, and *hydrogen fluoride.*

Froth Flotation. The separation of a *mixture* of finely divided minerals by agitating them in a froth of *water* and a frothing agent, so that some float and others sink. The process can be made selective by adjusting the nature of the froth with suitable *surface active agents.*

Fructose. Fruit sugar laevulose, $C_6H_{12}O_6$. A sweet *soluble* crystalline *hexose*, m.p. 102°C-104°C. It occurs in sweet ripe fruits, in the nectar of flowers, and in honey.

Frustum. Any part of a solid figure which is cut off by a *plane* parallel to the base, or lying between two parallel planes.

Fuel. Any material that can be burnt to supply heat or power.

Fulcrum. Pivot point on which a lever turns.

Fulminate of Mercury Hg $(OCN)_2$. Explosive substance obtained by dissolving mercury in nitric acid and then adding alcohol.

Fumaric Acid ($C_4H_4O_4$). Colourless, crystalline solid soluble in water.

Fumigant. Volatile chemical used in gaseous form for fumigation purpose *e.g.*, carbon disulphide, chloropicrin, ethylene, methyl bromide etc.

Fumigate. Expose to fumes especially for purpose of disinfecting or combating pests.

Fungicide. Substance that can destroy or kill fungizineb, ziram etc.

Furfuryl Alcohol ($C_5H_6O_2$). Colourless poisonous liquid, boiling point 171°C; obtained by reduction of furfural with sodium hydroxide.

Fuse. To melt. Metal safety device having low melting point, which breaks electric current when it becomes excessive or overloaded.

Fusible Metals. Alloys capable of melting at low temperatures *e.g.* alloy of bismuth, cadmium, lead and tin.

Fusion Mixture. Mixture of anhydrous potassium (K_2CO_3) and sodium carbonate (Na_2CO_3).

G

G. The symbol used for giga-or sometimes for gigabyte.

g. Stands for acceleration due to gravity.

GABA (Gamma amino-butyric acid). It acts as inhibitory transmitter in vertebrate nervous system.

Gadolinium (Gd). Silvery-white metal, malleable and ductile, boiling point 3266°C; used in electronic industry.

Galactose CH_2OH $(CHOH)_4$. CHO. A *hexose sugar*, m.p. 166°C; it is a *stereoisomer* of *glucose.* It is a constituent of *lactose* and certain plant *polysaccharides.*

Galaxies. Extra-galactic nebulae. The *stars* of the *Universe* are not evenly distributed throughout *space*, but are collected by gravitational attraction into some 10^9 giant clusters called galaxies. Each galaxy contains about 10^{11} stars. The *Sun* is one of such a number of a stars in our own *Galaxy*, which is itself a member of a *local group of galaxies.* The galaxies are separated from each other by enormous distances, the nearest galaxy the Milky Way being some 16×10^5 *light-years* away. Galaxies are either elliptical or spiral shaped; a very few however appear to have no regular shape.

Galaxy. The *Milky Way.* A cluster of some 10^{11} *stars*, one of which is the *Sun.* The Galaxy is a flat disc-shaped spiral structure, approximately 10^5 *light-years* across, with a slight bulge at the centre. The *solar system* is situated quite close to the central plane of this disc at a distance of about three fifths of its radius from the centre.

Galena. Natural lead (II) sulphide, PbS. A heavy crystalline *mineral* it is the principal *ore* of lead. It is used as a *semiconductor* in *crystal rectifiers.*

Galilean Telescope. Telescope invented by Galileo Galilei (1564-1642) but no longer used in *astronomy* although its principle is still used in opera glasses. It consists of a *bi-convex objective* of long *focal length* and a *bi-concave eye-piece* of short focal length, an erect image is produced.

Galactose ($C_6H_{12}O_6$). A sugar obtained on hydrolysis of lactose.

Gallic Acid ($C_7H_6O_5$). Colourless crystals having one molecule of water, sparingly soluble in water and alcohol; obtained from tannins by fermentation or acid hydrolysis; used in manufacture of inks.

Gallium (Ga). Silvery-white metal, boiling point 2403°C; used in lamps of spectroscope analysis.

Gallon. A unit of volume or capacity. The British Imperial gallon is the *volume* occupied by ten pounds of *distilled water*. It is equal to 4.54609 cubic decimetres (*litres*). The U.S. gallon is 0.83268 British gallons.

Galvanic Cell. *Primary cell.*

Galvanized Iron. Sheet iron coated with a thin layer of zinc to prevent *corrosion.*

Galvanometer. An instrument for detecting comparing, or measuring small *electric currents*, it requires calibration when an actual current measurement is needed. Galvanometers usually depend upon the magnetic effect produced by an electric current.

Gamboge. A yellow substance obtained from the hardened gumresin of the tree Garcinia Hanburii. It is used as a *pigment* and for colouring varnishes.

Gamma-rays. Electromagnetic radiation of very short wave-length (10^{-12} cm) very high energy.

Gangue. The useless stony minerals occuring with a metallic *ore.*

Garnet. A group of *minerals* of varying composition, mainly double *silicates* of calcium or aluminium with other *metals*. Several varieties are red in colour, and are used as gems.

Gas. A substance whose *physical state* (the gaseous state) is such that it always occupies the whole of the space in which it is contained. In a *perfect* gas, the atoms and molecules can move freely, have negligible *volume*, and have *elastic* collisions with each other.

Gas Carbon. Retort carbon. A hard deposit made up of fairly pure carbon, found on the walls of the *retorts* used for the *destructive distillation* of coal in the manufacture of *coal-gas*. It is a good *conductor* of *electricity*, it is used for making carbon *electrode.*

Gas Cleaning. Process of removing pollutants or contaminants from gases.

Gas-condenser. Apparatus used for freeing coal-gas from tar.

Gas-jar. Jar, usually of glass, used for collecting and holding gas in various chemical experiments.

Gas-light. Light produced by combustion of gas.

Gas Liquor. Solution of ammonia and ammonium salts prepared in process of gas-making.

Gas Mask. Respiratory device or breathing apparatus that covers mouth, nose and eyes of the person as protection against harmful.

Gas-meter. Instrument (meter) for measuring gas consumed.

Gasohol. Mixture of gasolene and 10 to 15% ethyl alcohol; used as fuel in interal combustion engines.

Gas Oil. Liquid petroleum distillate having viscosity and boiling range between kerosene and lubricating oil, includes diesel fuel, heating oils and light-fuel oils.

Gasolene (gasoline). Low-boiling petroleum fraction.

Gasometer. Very large iron vessel (reservoir) for storing gas; gas-tank.

Gas-stove. Apparatus in which coal-gas is used for heating or cooking.

Gas Constant. *R*. In the *gas equation, PV=RT*, the gas constant, *R*, equal to 8.31434 *joules* per kelvin per *mole* or 1.9858 *calories* per degree *celsius* per mole.

Gas-cooled Reactor. A *nuclear reactor* in which the *coolant* is a *gas*. In the *magnox* type, natural uranium *fuel* is used with a *graphite moderator*, the *fuel elements* being cased in magnox. The coolant is *carbon dioxide* and the outlet *temperature* is about 350°C.

Gas Turbine. An engine that converts the *chemical energy* of a *liquid fuel* into *mechanical energy* by internal *combustion*; the gaseous products of the fuel, are burnt in compressed air and then expanded through a *turbine*. It is used as the power plant in aeroplanes locomotives, and in motor vehicles. It is used as an auxiliary power plant in electrical generating stations.

Gate. **1.** A *circuit* having only one output. It has more than one input and can be activated by various combinations of input signals. **2.** A signal that activates a circuit for a predetermined time. **3.** A device for selecting portions of a *wave*, either on a time or an *amplitude* basis. **4.** The electrode in a *field-effect transistor*, it controls the flow of current through the channel.

Gauge Theories. Mathematical descriptions which predict the behaviour of all the *fundamental interactions*, except *gravitation*. These are based on the concept of *symmetry*. The *quantum theory* applied to the *electromagnetic interaction*, known as *quantum electrodynamics*, is a gauge theory in which the electromagnetic force is thought of as a result of the exchange of massless *photons* between charged particles. The success of this theory enabled it to be extended to encompass the *weak interaction* and to lead to the formulation of the gauge theory applying to *strong interactions*, known as *quantum chromodynamics*.

Gauss. The *c.g.s. system* unit of *magnetic flux density*. If a *magnetic field* of 1 *oersted* intensity exists in a medium of unit *magnetic permeability, e.g.*, air, then the induction will be 1 gauss. 1 gauss is equal to 1 *maxwell* per square centimetre or 10^{-4} *tesla*.

Gaussmeter. A *magnetomer* calibrated in *gauss*. It is used for measuring *magnetic flux density*.

Gauss's Law. The total *electric flux* of a closed surface in an *electric field* is proportional to the *electric charge* within that surface.

Gay-Lussac's Law of Gaseòus Combination. When *gases* combine, they do so in simple ratio by *volume* to each other, and to the gaseous product all measured under the similar conditions of *temperature* and *pressure*.

Gegenschein. Counter glow. A faint elliptical patch of *light* in the sky that may be observed at night directly opposite the *Sun*, though it is rarely seen in Britain. It is caused by a reflection of sunlight by meteoric particles in *space*.

Gelation. 1. The process of *freezing*. **2.** The formation of a *gel*.

Gelignite. An *explosive*. It consists of a *mixture* of *nitroglycerin, cellulose nitrate, sodium* or *potassium nitrate*, KNO_3, and wood pulp.

Generation Time. The average lapse of time between the creation of a *neutron* by *nuclear fission* and a subsequent fission produced by that neutron.

Generator. It is a device used for producing *electrical energy* from *mechanical energy*.

Generatrix (math.). A point, line, or *plane* the movement of which produces a line, surface, or solid.

Genetic Code. The code by which inherited characteristics are handed from generation to generation. The code is expressed by the molecular configuration of the *chromosomes* of *cells*.

Genetic Engineering. The technology involved in altering the genetic constitution of an organism. 'Useful' *genes*, are isolated from one organism and inserted into the DNA of a *bacterium* or *yeast*. These *microorganisms* multiply rapidly and can be cultured easily, enabling large quantities of the gene product to be obtained. Genetic engineering has been used for the large-scale production of *antibiotics, enzymes*, and *hormones (e.g. insulin).*

Genetics. The study of heredity in living organisms. The science has expanded to include the subdisciplines of microbial genetics, cytogenetics, and population genetics and has important applications in medicine, agriculture, and industry as well as in all aspects of *biology.*

Gentian Violet. A purple *dye* obtained from *anilline*, used as an *indicator*, an *antiseptic*, and a dye.

Geocentric. Measured from the centre of the Earth.

Geochemistry. The study of the chemical composition of the *Earth's* crust and the changes taking place within it.

Geodesic. Refers to the *geometry* of curved surfaces. A 'geodesic line' also called a 'geodesic', is the shortest distance between two points on a curved surface.

Geodesy. Surveying on such a scale that involves making allowance for the curvature of the *Earth.*

Geography. The study of the Earth's surface and the way man has interacted with it.

Geological Time Scale. Geological periods. A scale of time that serves as a reference for correlating various events in the history of the *Earth*; The time scale is divided into three main 'eras', based upon the general character of the life that they contain, each era being subdivided into 'period'.

Geology. The scientific study of the *Earth's* crust.

Geomagnetism. The study of the *magnetic field* associated with the *Earth.*

Germ. Micro-organism especially pathogen.

Germane (GeH_4). Colourless gas, germanium hyride.

Germanium (Ge). Rare metal, and is more reactive than silicon; used in alloys, glass, semiconductors etc.

German silver. Alloy of copper, nickel and zinc (2 : 1 : 1 or 5 : 2 : 2).

Germicide. Chemical used for killing or destroying germs.

Getter. Vacuum getter. A substance used for removing the last races of air or other *gases* in attaining a high *vacuum. e.g.*, magnesium metal is used in *thermionic valves.*

GeV. Abbreviation for *giga electron-volt, i.e.* 10^9 electron-volts. In America this is usually written BeV where the 'B' represents the American *billion.*

Ghosts (phys.) False lines that appear in a *line spectrum* due to imperfections in the ruling of the *diffraction grating* used.

Gibberellic acid (GA_3). Plant growth substance (gibberellin). It can be obtained from fungus *Gibberella fujikurai.* It causes disease of rice seedlings characterised by excessive elongation of shoots and leaves.

Gibberellines. A group of *plant hormones* that promote the growth of plant stems and fruit and have other beneficial effects.

Gibbous. Refers to the shape of the *Moon* or a *planet* when it is more than half-phase, but less than full-phase.

Gibbs' Function. *Free energy.*

Giga. Prefix denoting a thousand million (10^9). Symbol G.

Gilbret. The *c.g.s., unit* of *magnetomotive force* in *electromagnetic units.* It is equal to $10/4\pi$ *ampere-turns.*

Gilding. Covering with a thin layer of metallic gold, often by *electrolysis.*

Glacial Acid. Pure *ethonoic acid; solid* crystalline ethanoic acid below its *freezing point* (16.6°C.).

Glass. A hard brittle *amorphous mixture*, usually *transparent* or *translucent*, of the *silicates* of calcium, sodium or other *metals.*

Glass-ceramics. Materials usually consisting of lithium and magnesium *aluminosilicates*; they are chemically similar to glasses, but differ from glasses in consisting of very small crystals. They have high mechanical strength and very low thermal expansion, making them resistant to abrupt temperature changes. They are used for heat-resisting ovenware, for radomes, and other purposes involving exposure to drastic conditions.

Glass Fibre Materials. Fibreglass. Fine *glass* fibres, that are woven into a cloth and impregnated with various *resins*. They have high *tensile strength* and *corrosion* resistance, they are used in small boat buiding and for some motor vehicle body parts.

Glass Transition. The change, characteristic of many *rubbers* and other *polymers*, from a plastic or rubbery to a glassy or brittle state.

Glass-wool. Synthetic material resembling cotton-wool, comprising of very fine glass threads; used for filtering and absoring corrosive liquids.

Glauber's salt ($Na_2SO_4.IOH_2O$). Crystalline sodium sulphate.

Glaze. A *vitreous* covering for pottery.

Globular Clusters. Self-contained, approximately spherical, clusters of about one hundred thousand stars; some hundred of these clusters are known to be distributed about the centre of the *Milky Way*. They appear to be outside the *Galaxy*, they are believed to be gravitationally associated with it.

Globulins. Groups of *proteins soluble* in dilute solutions of mineral *salts*.

Glucagon. Proteinaceous hormone antagonistic to insulin in action. It converts glycogen into glucose in liver, it raises blood sugar level.

Glucan. Any polysaccharide made up exclusively of glucose sub-units *i.e.*, starch and cellulose.

Glucokinase. Enzyme. It causes addition of phosphoric acid radical to glucose.

Glucosuria (glycosuria). Refers to presence of glucose in urine.

Glucuronic Acid. Uronic acid derived from glucose; common constituent of gums and mucilages.

Glutamic Acid. Colourless, crystalline amino acid; used for flavouring purpose.

Gluten. Mixture of gliadin and glutenin that remains when starch of wheat grains has been removed.

Glycerine. Trihydric alcohol, colourless, sweet tasting, soluble in alcohol and water.

Glycine. Amino acids, polypeptides and proteins.

Glycogen. Animal starch, found in liver, on hydrolysis it yields glycose (glucose).

Glycoside. Compound formed by the action of pyranose sugar with a non-sugar molecule.

Glyoxsome. Microbody containing five enzymes of glyoxylate cycle.

Gold (Au). Bright yellow soft metal extremely malleable and ductile. It is not attacked by most acids, dissolves in aqua ragia; used in dentistry, jewellery, alloys etc.

Gold (III) chloride. Auric chloride. $AuCl_3$. A red *soluble* crystalline solid, that decomposes at 254°C. It is used in *photography* and gilding *glass*.

Gold Leaf. Gold is the most malleable of *metals*, and can be beaten into leaves 0.0001 mm thick. **The leaf** has the appearance of metallic gold, but transmits green *light; i.e.*, appears green when held up to the light.

Goldschmidt Process. The preparation of *metals* from their *oxides* by *aluminothermic reduction. i.e.*, by reducing the oxide by means of aluminium powder, *e.g.* $Cr_2O_3 + 2Al = 2Cr + Al_2O_3$.

Goniometer. Instrument used for measurement of angles between crystal faces.

Gout. It is a disease in which excess of uric acid in blood is deposited as sodium biurate in the joints etc., accompanied by swelling.

Gouy Balance. Balance used for determination of magnetic susceptibility.

Governor. Device for regulating speed of engine or machine on principle of negative feedback so that its speed is kept within check inspite of varying resistance.

Grahm's Law. Velocity of diffusion of gas in inversely proportional to square root of its density.

Gram (gramme :g). Unit of mass in metric system ($1g = 10^{-3}$ kilogram).

Gram-ion. Sum of atomic weight of an ion, expressed in grams.

Gramophone. Contrivance or play-box. It can reproduce sound, by playing recorded discs with help of small organ called sound box.

Grame-weight. Unit of force; pull of earth on one gram mass.

Granite. Coarse-ground igneous crystalline rock. It is composed of felspar, mica and quartz.

Graph. A diagram, generaly plotted between axes at right angles to each other, showing the relation of one *variable* quantity to another.

Graph. A suffix applied to instruments that automatically record or write down observations; *e.g. barograph.*

Graphite. Blacklead, plumbago. A natural *allotropic form* of carbon, used for pencil leads, in electrical apparatus, and as a lubricant for heavy machinery. It is also used as a *moderator* in *nuclear reactors.*

Graticule. Network of fine wires in eye-piece of microscope or telescope.

Gravimeter. Kind of hydrometer. It is for recording measurement under water and to determine presence of oil deposits under water.

Gravimetric Analysis. Branch of quantitative chemical analysis; amount of substance present is determined by converting it into some other substance of known chemical composition, which can be readily isolated, purified and weighed.

Gravitation. Mutual attraction between all bodies in universe.

Gravitational Force. Force on a body due to its gravitational attraction on other bodies.

Gravity. Gravitational attraction at surface of a planet or other celestial body.

Gray. The derived *SI unit* of absorbed *dose* of ionizing radiation. The energy in *joules* absorbed by one *kilogram* of irradiated material.

Grease. A semi-solid lubricant consisting of emulsified *petroleum* oils and *soluble hydrocarbon soaps.*

Greate Circle. A circle obtained by cutting a *sphere* by a *plane* passing through the centre. *e.g.,* regarding the Earth as a sphere, the *Equator* is a great circle, as are all the *meridians* of longitude.

Greek Fire. A mixture of material that caught fire when wetted; used by the ancient Greeks in naval warfare. It is composed of sulphur, *naphtha,* and *calcium oxide* or similar materials.

Green. Hue, evoked in an average observer by monochromatic radiation having wave-length in range from 492 to 577 nanometres.

Green Vitroil. Ferrous sulphate crystals. ($FeSO_4 . 7H_2O$).

Gregorian Telescope. A form of astronomical reflecting *telescope.* Although this type of telescope gives an erect image it is not now widely used as it is difficult to adjust.

Grid. A system of *high tension* cables (overhead or underground) by which electrical power is distributed throughout a country. A network of horizontal and vertical lines superimposed on a map to enable references.

Grid Bias. A fixed *voltage* applied between the *cathode* and the *control grid* of a *thermionic valve,* which determines its operating conditions.

Grignard Reagents. *Alkyl* magnesium *halides,* prepared by the action of magnesium metal on alkyl halides in *ether* solution; *e.g.* C_2H_5MgI.

Griess Reagent. Solution of sulphanilic acid, α-naphthyl-amine and acetic acid in water; used to test for nitrous acid.

Ground Wave (surface waves). Electromagnetic radiation of radio frequencies that travel directly from transmitting aerial to receiving aerial.

Guided Missiles. Rocket propelled missiles that are controlled during flight by self-activating or by radio signals from the ground.

Gun-cotton. Explosive. It can be prepared by saturating cotton-wool with nitric acid and sulphuric acid.

Gun-metal. Alloy of copper and tin (9 : 1) contains upto 4% zinc; used for making cannon-barrels.

Gun Powder. Explosive mixture consisting of charcoal, sulphur and potassium nitrate.

Gyro-compass. Compass that does not make use of magnetism and is therefore not effected by magnetic storms.

Gyromagnetic Ratio. Ratio of magnetic moment of an atom or nucleus to its angular momentum.

Gyroscope. Apparatus in which heavy fly-wheel rotates at high speed to illustrate dynamics of rotating bodies; used as educational toys, stabiliser of ships etc.

H

Heber Process. Industrial process for manufacturer of NH3 in which direct combination of nitrogen and hydrogren takes place under high pressure (250 atm.) and low temperature (450°C); ammonia gas formed is condensed to liquid and unreacted gases are recirculated.

Habitat. The place where an animal or plant lives, the organism being adopted to the particular environmental conditions within the habitat, which may be a sea-shore, pond, etc.

Hadron. Any elementary particle which has strong interaction-baryons, mesons etc.

Haematin. Brownish substance, contains ferric ion obtained from oxyhaemoglobin or from dried blood.

Haematite. Iron ore, Fe_2, O_3, red in colour.

Haeme. Red pigment combined with protein (globin) in haemoglobin.

Haemianopsia. Half-blindness.

Haemoglobin. Red oxygen carrying pigments present in red blood corpuscles.

Haemophilia. Inherited tendency to excessive bleeding even when any blood vessel is slightly injured; six-linked disease common in males.

Haemostat. Instrument used for stopping bleeding.

Hafnium (Hf). White metal, burns in air and forms HfO_2, boiling point 5230°C; used in surgery and in making wireless valves.

Hair Salt. Natural aluminimium sulphate $Al_2(SO_4)_3 \cdot 18H_2O$.

Half life. Time in which half of radio active isotope in any given sample disintegrates.

Half -period Zones. The division of a *wave front* into elements of area or zones such that secondary wavelets reaching a given point ahead of the wave from adjacent zones differ in *phase* by half a period, or π. This construction is used in theoretical investigations of *Fresnel diffraction* in simple cases.

Half-Thickness. The thickness of a specified material that, when introduced into the path of a given beam of *radiation* reduces its intensity to one half of its original value.

Half-wave Plate. A plate of double refracting material cut parallel to the *optic axis* and of such a thickness that a *phase* difference of π or 180° is introduced between the *ordinary ray* and the extraordinary ray for *light* of a particular *wavelength.* It is chiefly used to alter the plane of vibration of plane-polarized light.

Halide. Binary compound containing one of the halogen atoms.

Hall Effect. On passing an *electric current* in a wire placed in a strong transverse *magnetic field,* a *potential-difference* is developed across the wire, at right angles to both the magnetic field and the wire.

Halley's Comet. A bright *comet.* It takes about 76 years to orbit the *Sun;* last seen in 1986. It moves round the Sun in the opposite direction to the planets.

Hall-Heroult Cell. An *electrolytic cell* used for extracting aluminium from *bauxite* ($Al_2O_3xH_2O$). After purification the oxide is mixed with *cryolite* (Na_3AlF_6) so as to lower its melting point; this forms the *electrolyte,* which is maintained at about $850^{\circ}C$. by means of the current passed through it between the *graphite* cell lining, which functions as a *cathode,* and the graphite *anodes.* Molten aluminium is tapped off from the base of the cell.

Hall Mobility. The mobility of *carriers* in a *semiconductor;* it is the *velocity* of the carriers under the influence of an *electric field* of 1 *volt per metre.*

Hallucinogen. Any substance that induces hallucinations.

Halo. Luminous ring observed surrounding sun or moon; It is caused by refraction of light by ice crystals in atmosphere.

Halobiontic. Strictly confined to salt water.

Halogen(s). Refers to astatine, bromine, chlorine, flourine and iodine—having closely related and graded properties.

Halogenation. Process of introduction of halogen atoms into compound by addition or substitution.

Halophyte. Plant that grows in soils having high concentration of salts.

Hard Glass. Glass containing high amount of potassium and silica; It is used for making glassware.

Hard Iron. Iron which is not easily magnetised by induction.

Hard Radiation. Radiation having particles of photons which have high energy and can penetrate through a large number of materials including metals.

Hard Water. Water that does not easily lather with soap. It is due to presence of calcium, magnesium and iron compounds in that water.

Harmonic. Series of sounds, each of which has frequency that is an integral multiple of some fundamental frequency.

Harmonic Series (math.). A *series* in which the *reciprocals* of the terms are in *arithmetical progression, e.g.*

$$1 + {}^1/_2 + {}^1/_3 + {}^1/_4 \ldots$$

Harmonics of a Wave Motion. Waves superimposed on a fundamental wave, having a *frequency* that is a whole multiple of the fundamental frequency. The second harmonic has a frequency twice that of the fundamental, the third harmonic three times, and so on.

Hartree. Atomic unit of energy. A unit of energy equal to e^2/a_o, where e is the charge on an electron and a_o is the atomic unit of length. It is equal to 4.85×10^{-18} joule or 27.2 eV.

Hartsthorn, Spirits of. A *solution* of *ammonia in water.*

Heat. Energy possessed by substance as kinetic energy of atomic or molecular translation, rotation or vibration; It is expressed in calories or joules.

Heat Capacity. It is the amount of heat (joules) required to raise temperature of a body through 1°C.

Heat death of the Universe. According to the second law of thermodynamics the *entropy* of a closed system tends towards a maximum and that its available *energy* tends towards a minimum. The *Universe* constitutes a thermodynamically closed system, it would mean that some time no energy will be available for use. This state is referred to as the 'heat death of the Universe'.

Heat Engine. Any device that makes use of *heat* to do *work.*

Heat Exchanger. Any device that transfers *heat* from one *fluid* to another without allowing the fluids to come into contact with each other. The simplest type consists of a cylinder within which a coiled tube is mounted. One fluid passes through the coiled tube in one direction, while the other fluid passes through the cylinder, outside the tube, in the other direction.

Heat Flux. Amount of heat transferred across a surface of unit area in a unit time.

Heat of Adsorption. Increase in enthalpy when one mole of substance is adsorbed over another at constant pressure.

Heat of Atomisation. Heat required to dissociate one mole of substance into atoms.

Heat of Combustion. Amount of heat liberated when one mole of substance completely burns in excess oxygen.

Heat of Crystallisation. Quantity of heat evolved when a mass of liquid crystalises at its freezing point.

Heat of Decomposition. Change in enthalpy during decomposition of one mole of compound into its elements at constant pressure.

Heat of Dilution. Increase in enthalpy accompanying addition of any specified amount of solvent at constant pressure.

Heat of Dissociation. Increase in enthalpy when molecules break at constant pressure.

Heat of Formation. Amount of heat required for formation of one mole of substance from its elements at constant pressure.

Heat of Fusion. Amount of heat required to convert a unit mass of solid to liquid at melting point.

Heat of Neutralisation. Heat given out when one mole of acid or base is neutralised.

Heat of Reaction. Amount of heat given out or absorbed in a chemical reaction.

Heat of Solution. Amount of heat changes accompanying dissolution of one mole of substance in large volume of water or any other solvent at constant pressure.

Heat of Sublimation. Quantity of heat required to sublime a unit weight of solid.

Heat of Vapourisation. Heat required to vapourise a unit mass of given liquid at constant temperature.

Heat Shield. The shielding surface or structure that protects a *spacecraft* from excessive heating on entering the *Earth's* atmosphere.

Heat Transfer. The transfer of *energy* from one body or place to another as a result of a difference in *temperature.*

Heaviside-Kennelly layer. A region of the *ionosphere* lying between 90 and 150 kilometers above the surface of the *Earth* that reflects *electromagnetic radiation of radio frequencies.*

Heavy Hydrogen. *(D) deuterium.*

Heavy Metal. A metallic *element* having relatively high *relative atomic mass, e.g.*, platinum, gold, lead.

Heavy Water (Deuterium oxide D_2O). Water in which hydrogen is replaced by deuterium $H_2O + D_2 \rightleftharpoons D_2O + H_2$; used as moderator in nuclear reactors, also as tracer in metabolic bolic studies.

Helium (He). Light mono-atomic gas obtained from liquid air or natural gas, boiling point 269°C; used for filling air-ships and baloons.

Henry's Law. Solubility of gas in liquid is directly proportional to pressure of gas at constant temperature.

Heparin. Organic acid. It contains N and S, prevents blood cloting.

Hepatic. Relating to liver.

Hepta. Prefix, meaning seven.

Heptachlor. ($C_{10}H_5Cl_7$); white crystalline solid, insoluble in water, soluble in alcohol, k. oil. Stable and used as non-systemic, stomach and contact insecticide.

Heptavalent. Having a valency of 7.

Herbicide. Any substance that kills or inhibits the growth of plants.

Herbivore. Refers to animal that eats only grasses or plants.

Heredity. Sum of genetic endowment obtained from parents.

Herparin. Substance that prevents blood clotting by neutralising prothrombin.

Hertz (hz). S.I. (System International) unit of fequency (1 hz = one cycle per second).

Hess's Law. Change in ethalpy for any chemical reaction is constant, whether reaction occurs in one step or in many steps.

Heterogeneous. Non uniform; showing different properties in different portions.

Heterogeneous Reaction. Reaction which takes place between substances of different phases *e.g.*, between liquid and gas.

Heuristic. The method of solving mathematical problems for which no *algorithm* exists; it involves the narrowing down of the field of search for a solution by inductive reasoning from past experience of similar problems.

Heusler's alloys. *Alloys* containing neither iron, nor cobalt that exhibit strong *ferromagnetism*. They are composed of copper, manganese and aluminium.

Hexa. Prefix denoting six.

Hexact. Six-rayed sponge spicule.

Hexad. Atom, element or radical having a combining power of six.

Hexadactylons. Six-fingered; six-toed.

Hexadecane ($C_{16}H_{34}$). *Cetane*; Colourless liquid, boiling point 287°C; used as solvent and in determination of cetane numbers.

Hexadecanoic Acid. *Palmitic acid.*

Hexadecanol. Cetyl alcohol. $C_{16}H_{33}OH$. White crystalline *insoluble solid*, m.p. 50°C., used in cosmetics and in pharmaceutical products.

Hexamine. Hexamethylenetetramine, urotropine. $(CH_2)_6N_4$. White crystalline substance obtained by *condensation* of *ammonia* with methanal. Used in medicine, in the manufacture of vulcanized *rubber*, and in the manufacture of *cyclonite.*

Hexane (C_6H_{14}). Sixth member of methane series; found in petroleum.

Hexanoic Acid [$CH_3(CH_2)_4$ COOG]. Colourless oily liquid with unpleasant small, boiling point 205°C; used in artificial flavouring.

Hexavalent. Having valence of six.

Hexene (V_6H_{12}). Unsaturated hydrocarbon of ethylene series.

Hexkinase. Enzyme which catalyses phosphorylation of hexose sugars like glucose.

Hexoson (s). Polysaccharide which yields hexoses on hydrolysis.

Hexose. Monosaccharide (sugar), its molecules contains six carbon atoms *e.g.*, galactose, glucose, fructose etc.

Hexylresorcinol [$C_6H_{13}C_6H_3(OH)_2$]. Yellow crystalline solid; used as an antiseptic and in medicines.

High Frequency. Ratio frequencies between 3,000 and 30,000 kilohertz.

Histamine. $C_3H_3N_2$ $(CH_2)_2$ NH_2. White crystalline substance, m.p. 86°C. It occurs in animal *tissues* when they are injured and as part of the allergic reaction, causing dilation of blood vessels; stimulates gastric secretion of *hydrochloric acid.*

Homocentric. Those rays that have same focal point or are parallel.

Homochromous. That are alike in colour.

Homochromy. Protective colouration.

Homocyclic. Having a close chain of like atoms.

Homodesmic. Crystal that has atoms bonded in a single way.

Homogeneous. Refers to same kind or nature; having constituent elements similar throughout.

Homogeneous Radiation. Radiation that have extremely narrow band of frequencies.

Homologous Series. Chemical compounds of uniform type. They show regular gradation in physical properties.

Homologue(s). Homologous to something or other *e.g.*, human arm, bird's wing and whale's flipper. Members of some homologous series *e.g.*, ethane (C_2H_6) and methane (CH_4).

Homology. Affinity of structure and origin, apart from and use.

Homolysis. Breaking of chemical bond in such a way that neutral atoms or radicals are formed.

Homomorphous. Molecules that are similar in shape and size but not necessarily in other characters. They have same crystalline form but different chemical composition.

Honey. Sweet viscous fluid elaborated by bees from nectar of flowers.

Honey-bag. Enlargement of alimentary canal of bees in which honey is stored.

Honey-buzzard Falcon (*Pernis apivona*) that feeds on bees, wasps etc.

Honeycomb. Structure of hexagonal waxy cells constructed by bees, for storing honey

Hooks's Law. Within the elastic *limit*, a *strain* is proportional to the *stress* producing it.

Hormone(s). Natural organic substances, they regulate growth metabolism and other functions of an organism.

Hornblende. A rock-forming *mineral.* It consists mainly of *silicates* of calcium magnesium, and iron.

Horn Silver. Natural *silver chloride*, AgCl. An important *ore* of silver.

Horse Power. British Unit of power equal to work of 550 foot poundals/Sec. 1HP =745.7 watts.

Hot-wire Instrument. An electrical instrument (*ammeter or voltmeter*). It depends upon the expansion, or change in *resistance*, of a wire heated by the passage of an *electric current.* It will measure both a.c. and d.c.

Hovercraft. Vehicle (aeroplane type) that, instead of moving on wheels propels itself on a cushion of air formed by valves on its underside and can move on both land and water, a few inches above the surface.

Hubble Constant. The ratio of the distance between the *Local Group of galaxies* and receding cluster of glaxies to the rate at which the distant cluster recedes. The value of the Hubble constant is estimated as being between 5 and 10 thousand million years.

Hue. Characteristic of colour such as blue, red etc. It is determined by its wave-length.

Humidifier. A device for supplying moisture to air so as to maintain desired humidity conditions.

Humidity. Amount of water vapour present in air.

Humus. Dark brown colloidal matter present in soil due to decomposition of microbial, animal and plant materials.

Hyaline. Glassy or semitransparent.

Hybrid. Individual resulting from union of egg and sperm that differ in one or more genes; offspring of two different species *e.g.*, mongrel, mule.

Hydragogue. Removing water or serum; drug having that effect.

Hydragyrisms. Mercurial poisoning.

Hydrazide. Organic compound having general formula.

$$RCO—NH—NH_2.$$

Hydride. Binary compound with hydrogen.

Hydrobromic Acid (HBr). Colourless, pungent smelling, highly soluble gas can be obtained by hydrolysis of phosphorus bromide; used as laboratory reagent in organic chemistry.

Hydrocarbons. Compounds containing carbon and hydrogen only Hydrocarbons may be—

— **saturated,** four valencies of each carbon atom satisfied by a single bond *e.g.*, ethane (C_2H_6), methane (CH_4) etc.

— **unsaturated,** valencies of atleast two adjoining carbon atoms not fully satisfied by single covalent bond *e.g.*, ethene ($CH_2=CH_2$), ethyne ($OH=CH$).

— **aromatic,** they have ring structure-benzene (C_6H_6), naphthalene ($C_{10}H_8$), toluene (C_7H_8) etc.

Hydrochloric Acid (HCI). Colourless, pungent smelling, highly soluble gas, can be obtained by heating sodium chloride with concentrated sulphuric acid; used as laboratory reagent.

Hydrochloride. Salt formed when an organic base combines with hydrochloric acid.

Hydrocracking. Cracking of petroleum or its products in presence of hydrogen and a suitable catalyst.

Hydrodistillation. Process of removal of essential oils from plant.

Hydroelectricity. Electric energy obtained by using water-power to drive the dynamo.

Hydrofluoric Acid (HF). Colourless, highly poisonous liquid can be obtained by heating calcium fluoride with concentrated sulphuric acid; very corrosive to skin; used as antiseptic by brewing industries.

Hydrogel. Colloidal gel in which water acts as the dispersion medium.

Hydrogen (H). Gaseous substance, very light in weight, colourless, on combination with oxygen, it forms water (H_2O).

Hydrogenation. Subjecting to chemical action of or causing to combine with hydrogen.

Hydrogenation of Coal. The manufacture of artificial mineral *oil* from *coal* by action of hydrogen.

Hydrogenation of Oils. Artificial hardening of *liquid* animal and *vegetable oils* by the action of hydrogen. This is the basis of the process for manufacturing *margarine.*

Hydrogen Azide. Hydrazoic acid, azoimied. HN_3. A colourless poisonous *explosive liquid*, b.p. 37°C., it forms explosive *salts* with heavy *metals*. These are called *azides.*

Hydrogen Bond. A weak *elactrostatic* chemical *bond* that is formed between covalently bonded hydrogen *atoms* and a strongly *electronegative* atom with a *lone pair of electrons* (*e.g.*, oxygen, nitrogen, fluorine). *Ice crystals* are held together by this type of bond. When ice melts this structure breaks but some hydrogen bonds continue to exist, and liquid *water* consists of group of water *molecules* held together by hydrogen bonds. The hydrogen bond is of enormous importance in biochemical processes, especially the N–H - - - N bond, which enables complex *proteins* and *nucleic acids* to be built up. Life would be impossible without this type of linkage.

Hydrogen Bromide. *Hydrobromic acid.*

Hydrogen Carbonate Bicarbonate. A salt of carbonic acid in which one hydrogen atom has been replaced, *i.e.*, a salt containing the *ion* HCO_3^-.

Hydrogen Chloride. *Hydrochloric acid.*

Hydrogen Cyanide. *Hydrocyanic acid.*

Hydrogen Eletrode. A *half cell* used as a standard of measuring *electrode potentials*, it is assigned a potential of zero. It consists of a platinum electrode, over which hydrogen is bubbled, immersed in a dilute acid.

Hydrogen Fluoride. *Hydrofluoric acid.*

Hydrogen Iodide. *Hydroiodic acid.*

Hydrogen Ion. A positively charged hydrogen *atom*; a *proton.* The general properties of *acids* in *solution* are due to the presence of hydrogen ions.

Hydrogen Ion Concentration. The number of moles of *hydrogen ions* per cubic decimetre of *solution.*

Hydrogen Peroxide. H_2O_2. A thick syrupy *liquid*, b.p. 150.2°C; it is sold as a *solution* of the pure compound in *water*. It gives off oxygen readily, and is used as a *disinfectant* and *bleaching* agent. Strength of solution is usually given in terms of volume strength; thus 10 volume hydrogen peroxide will evolve 10 times its own volume of oxygen *gas* at N.T.P.

Hydrogen Phosphide. *Phosphine.*

Hydrogensulphate Bisulphate. A salt of sulphuric acid in which one hydrogen atom has been replaced, *i.e.*, a salt containing the ion HSO_4^-.

Hydrogen Sulphide H_2S. A colourless poisonous *gas* with a smell of rotten eggs.

Hydrogensulphite Bisulphite. A *salt* of *shlphurous acid* in which one hydrogen atom has been replaced, *i.e.*, a salt containing *ion* **$HSO3^-$**.

Hydroiodic Acid HI. A *solution* of the colourless *gas*, hydrogen iodide, in water.

Hydrolases. A class of *enzymes* that control *hydrolysis; e.g., esterases, proteases.*

Hydrolith. Calcium hydride. CaH_2. A substance that is *decomposed* by *water* and is used for the production of hydrogen, according to the following equation :

$$CaH_2 + 2H_2O = Ca(OH)_2 + 2H_2.$$

Hydrological Cycle. Water cycle. The cycle of events by which water evaporates from the seas into the atmosphere, and return back.

Hydrology. The study of *water* with reference to its occurrence and properties in the *hydrosphere* and *atmosphere.*

Hydrolysis. The chemical *decomposition* of a substance by *water*, the reaction is of the type :

$$AB+H_2O = A\,(OH) + HB.$$

Salts of weak *acids,* weak *bases,* or both are partially hydrolyzed in *solution, esters* may be hydrolyzed to form an *alcohol* and acid.

Hydrometer. An instrument used foι measuring the *density* or *relative density* of *liquids.*

Hydronium Ion. *Oxonium ion,* H_3O^+.

Hydrophilic. *Water loving.*

Hydrophobic. *Water-repellant.*

Hydroponics. Cultivation of plants without the use of *soil,* instead of soil *solution* of those mineral *salts* that a plant normally extracts from the soil are used.

Hydroquinone. *Benzene-1, 4. diol.*

Hydrosol. A *colloidal solution,* as distinct from a *hydrogel, water* being the *solvent.*

Hydrosphere. The watery portion of the *Earth's* crust, comprising the oceans. Seas, and all other waters.

Hydrostatics. The study of *forces* and *pressures* in *liquids.*

Hydrosulphate. A *salt* formed when an *organic base* (*e.g.,* an *alkaloid*) combines with *sulphuric acid.*

Hydrosulphide. *Hydrogen sulphide ion.*

Hydrosulphuric Acid. *Hydrogen sulphide.*

Hydrous. Containing *water.*

Hydroxide. A *compound* obtained from *water,* H_2O, by the replacement of one of the hydrogen *atoms* in the *molecule* by some other atom or group; either a compound containing the hydroxide ion, OH^-, or a compound containing the *hydroxyl group,* –OH. Hydroxides of metals are *alkalies.*

Hydroxy Acid. An *organic acid* that contains *hydroxyl* groups in addition to *carboxyl* in its *molecule; e.g., lactic acid.*

$$CH_3CH(OH)COOH.$$

5-Hydroxy-2-butanon. *Aceton.*

2-hydroxy-1, 2-diphenylethanone Benzo in. $C_6H_5CHOH.CO.C_6H_5$. An *optically active* crystalline substance, m.p. 133—7°C., used in organic synthesis.

Hydroxyl Group. The *univalent*—OH group. It is present in covalently bonded form in *alcohols.*

2-hydroxypropanoic Acid. *Lactic acid.*

Hydro. Denoting moisture, humidity, *e.g., hygrometer.*

Hydrodeik. A *wet* and *dry bulb hygrometer* with a chart attached. It enables the *relative humidity* to be obtained directly from the readings of the two *thermometers.*

Hygrometer. Instrument used for showing variations of *relative humidity* of the *atmosphere.*

Hygroscope. Instrument used for showing variation of *relative humidity* of the air.

Hygroscopic. That has a to tendency to absorb moisture.

Hyoscyamine ($C_{17}H_{23}No_3$). Poisonous crystalline substance obtained from henbane; used as sedative and antispasmodic.

Hyperacapina. Refers to excessive carbon dioxide in blood.

Hyperchromatism. Excessive pigments in skin.

Hypergolic. Which is capable of igniting spontaneously upon contact.

Hypermetropia. Defect of eye to see near objects clearly. It can be corrected by use of convex lens.

Hyperparasite. Parasite that parasitises another parasite.

Hypersonics. Production and utilisation of sound waves of frequencies above 500 megahertz.

Hypertension. High blood pressure.

Hypertonic. One of the two solutions that has greater osmotic pressure than the other.

Hypnotic. Sedative. Any substance that induces sleep.

Hypo ($Na_2S_2O_3.5H_2O$). Sodium thiosulphate; used in photography.

Hypochlorous Acid (HClO). Weak acid, prepared by reaction of chlorine on mercurous oxide; used as bleaching agent.

Hypochlorite. Salt of hypochlorous acid.

Hypophosphite. Salt of hypophosphorous acid.

Hypophosphorous Acid (H_3PO_3). Colourless, crystalline, soluble solid, known as phosphinic acid.

Hypothesis. Proposition assumed which is yet to be proved or disproved by facts.

Hypotonic. One of the two solutions that has lower osmotic pressure as compared to the other.

Hypoxanthine. Crystalline powder, slightly soluble in water, dissolves readily in alkalies; present in muscles, liver, spleen and pancreas.

Hypoxia. Detrimental effects caused due to lack of oxygen in body cells or tissues.

Hypsometer. Instrument used for measuring atmospheric pressure to ascertain elevations by determination of boiling point of liquid. It is also used for calibrating thermometers at steam temperature.

Hyssop. Aromatic, strong smelling plant (*Hyssopus officinalis*), formerly used in medicines but now only as ornamental plant.

I

IAA (inodole acetic acid). Principal auxin in plants. It is associated with a number of plant responses in very low concentrations.

IAR. Abbreviation for instruction address register.

Iatrochemistry. Refers to medieval medical chemistry introduced by Franciscus Sylvius (1614-72); confined mainly to application of drugs to medicine.

IC. Abbreviation for integrated circuit.

Ice. Crystalline form of water, melting point $0^\circ C$, latent heat of melting 80 calories per gram.

Ice nucleus. Any particle that can act as central point in formation of ice crystals in atmosphere.

Ice point. Equilibrium temperature between ice and water under normal atmospheric pressure ($0^\circ C$).

Iconoscope. Camera tube in which an *electron* beam scans a *mosaic*, this converts an optical image into an electrical signal.

Icosahedron. A *polyhedron* with twenty faces.

Ichthyosis. Disease in which skin becomes hard, rough and thick.

Icicle. Tapering piece of ice formed by freezing of dropping water.

Icosahedral. Solid figure with 12 apices, each being triangular.

Icterus. Pathological condition in which skin, scleres, mucous membrane etc., turn yellow due to accumulation of free or conjugated bile pigment bilirubin.

Ide. Suffix devoting binary compounds of two named elements e.g. hydrogen sulphide.

Ideal Crystal. Crystal whose lattice is perfectly regular containing no foreign atoms or ions.

Ideal Gas. Gas whose molecules are infinitely small exerting no force on each other.

Ideal Solution. Solution showing no change of internal energy on mixing and no attractive forces between components.

IDL (intermediate density lipoproteins). Substances formed during fat metabolism in liver by conversion of very low density lipoproteins by loss of some triglycerides and protein in the complexes.

Idle Time. Time during which the central processor of a computer system is performing no useful function.

Igneous Rock. A rock formed from molten silicates (*magma*).

Ignite. To heat to the point at which combustion takes place; to set on fire.

Ignition. Initiating combustion by increasing temperature of reactants to ignition temperature.

Illuminance E_υ. The *luminous flux* falling on a surface per second. *SI unit* of illuminance is the *lux* (*lumen* per square *metre*).

Ilmenite. $FeTiO_3$. An *ore* of *titanium.*

Imbibition. Phenomenon in which substance absorbs a liquid and so it swells.

Imide. Compound obtainted from ammonia in which two hydrogen atoms of ammonia are replaced by metal atoms or acid radicals.

Imine. Nitrogen containing organic compound having carbon-nitrogen double bond (R—HC=NH), these are formed by condensing aldehydes or ketones with ammonia.

Imino Acid. Organic acid in which=NH group is attached to one or two carbons.

Immiscible. Incapable of being mixed to form homogeneous mixture.

Immune. Not liable to catch infection.

Immune Haeolysin. Substance which is formed in blood in response to injection of erthocytes from another species.

Immune Protein. Refers to antibody.

Immune Serum. Blood serum obtained from an immunised individual carrying antibodies.

Immunisation. Process of making an organism that is resistant to infection or harmful agents.

Immunoglobin. Antibody protein generated to a specific antigen.

Impact. The collision of bodies.

Impedance Z. The quantity which determines the *amplitude* of the current for a given *voltage* in an *alternating current circuit.*

Imperial Units. British system of units based on the *pound, yard* and *gallon.*

Impermeable. Not permitting passage of fluids.

Implosion. Inward collapse of an evacuated vessel.

Impulse. Effect of force that is suddenly or momentarily communicated.

Impurity. Undesirable foreign material present in a pure substance.

Incandescence. Emission of light caused due to high temperature; white heat.

Inclinometer. An instrument for measuring the angle of inclination that an aircraft makes with the horizontal.

Incubator. A box designed to maintain a constant internal *temperature* using a *thermostat*; it is used for rearing chickens and prematurely born infants.

Indene ($C_6H_4C_3H_4$). Colourless liquid boiling point 182.5°C, obtained from coal-tar.

Index. Alphabetical list of names, subjects etc., along with page numbers where these appear in the text.

Indicator. Substance usually derived from plant material that can exist in either acid or base form— methyl orange, methyl violet etc.

Indigo (indigotin) ($C_{16}H_{10}O_2N_2$). Dark blue powder used mainly as dye-stuff. Indigo plant (*Indigofera spp.*).

Indigo Red. An isomer of indigotin that is obtained from natural indigo.

Indium (In). Soft silvery-white metal, boiling point 2080°C; used in electroplating and dentala lloys.

Indole (C_5H_7N). Yellow soluble substance. It occurs in oil of jasmin; used in perfumes.

Indole-3-acetic Acid. IAA, indole-3-ethanoic acid $C_{10}H_9NO_2$. A white crystalline substance, m.p. 168–170°C. It promotes plant growth .

Induced Radioactivity. Artificial radioactivity. *Radioactivity* induced in naturally stable *elements* by bombarding them with *neutrons* or other high *energy* particles (or *photons*).

Inductance. The property of an electric *circuit* as a result of which an *electromotive force* is generated by a change in the *current* flowing through the circuit or by a change in the current of a neighbouring circuit with which it is magnetically linked. The *SI unit* of inductance is the *henry.*

Induction Charging by. A process of electrically charging an insulated *conductor*, using the *force* due to another nearby charge to separate the positive and negative charges present on the conductor.

Induction, Electromagnetic. When the *magnetic flux* through a circuit changes, an electromotive force is induced in the circuit. This phenomenon is called induction.

Induction Coil. An instrument for producing a high *electromotive force* from a supply of low *E.M.F.* Induction coils are widely used in *internal- combustion engines* to provide the electric sparks.

Induction Heating. A form of heating in which electrically conducting material is heated as a result of the *eddy currents* induced in it by an alternating *magnetic field.*

Induction Motor. A type of *electric motor* in which an *alternating current* supply fed to the primary winding sets up a flux that causes *electrical currents* to be induced in the secondary winding of the *rotor*. The interaction between these currents and the flux causes the rotor to rotate.

Inductometer. A calibrated variable *inductance.*

Inelastic Collision. A collision between bodies in which there is no loss of total *energy*. Referring to *nuclear physics*, an inelastic collision is one in which an incoming particle causes *excitation* or breaking up of the struck *nucleus.*

Inert. Does not easily take part in *chemical reaction.*

Inert Gases. Noble gases, rare gases. The *elements helium*, neon, argon, krypton, xenon, radon forming group 0 of the *periodic table*. They are all chemically inactive.

Inertia (Phys.). The tendency of a body to maintain its state of rest or uniform motion in a straight line.

Inertial Guidance. A method of automatic control used in *guided missiles* that depends upon *inertia easily.*

Inertial Mass. The *mass* of a body as determined by its *momentum* as opposed to 'gravitational mass', which is determined by the extent to which it responds to the *force of gravity*. The *acceleration* of a falling body increases in proportion to its gravitational mass and decreases in proportion to its inertial mass. Since all falling bodies have the same constant acceleration it follows that the two types of mass must be equal.

Intertial System. A *frame of reference* in which bodies are not accelerated, *i.e.,* remain at rest or move with constant *velocity*, unless acted upon by external *forces.*

Infect. To contaminate; enter and establish pathogenic relationship with host.

Infest. To be present (usually in a large number) and cause inconvenience and/or injury to host.

Infinitesimal. A quantity smaller than any assignable quantity.

Infinity, ∞. A quantity that is greater than any assignable quantity.

Inflection. A point on a *curve* at which the *tangent* to the curve changes its direction of rotation. In the curve $y= f(x)$, a point of inflection has the characteristic that $d^2y/dx^2=0$.

Information Theory. Branch of *cybernetics* in which an attempt is made to define the amount of information required to controll a process of given complexity.

Infrared Radiation IR. *Electromagnetic radiation* possessing *wavelengths* between those of visible *light* and *radio* waves, *i.e.,* from approximately o.8μ m to 1 mm. Infrared radiation has the power of penetrating *fog* or haze, which would scatter ordinary visible light; thus photographs taken on a plate made sensitive to infrared radiation may often disclose detail invisible on an ordinary plate or to the naked eye.

Infrared Stars. Celestial bodies which emit *infrared radiation.* They consist of *stars* surrounded by dust clouds. In some cases the *light* from the central star penetrates the dust and it can be seen with *optical telescopes.*

Infrasonic. Having a *frequency* less than the frequency of audible *sound waves, i.e.*, a frequency of less than about 20 *hertz.*

Infrasound. Vibrations or pressure waves with a *frequency* less than that of *sound, i.e.*, below about 20 *hertz.*

Infusible. Difficult to melt, having very high melting point.

Innominate. Having no name.

Inorganic. Of mineral origin.

Inorganic chemistry. Branch of chemistry that deals with study of metallic and non- metallic elements and their compounds.

Inositol [$C_6H_6(OH)_6$]. Optically active crystalline solid. It occurs in vitamin B-complex and is an essential component of animal diet.

Insecticide. Toxin substance effective against insects.

Insectivore. Insect eating organism.

Insolation. Exposure of sun's rays.

Insoluble. Not capable of forming a *solution* in *water.* It is a relative term, since most substances dissolve in water to some extent.

Instantaneous Frequency. The rate of change of *phase* of an oscillation.

Insulation. The prevention of the passage of an *electric* current, or *heat,* by *conduction.*

Insulator. A non-conductor of *electricity* or *heat.*

Insulin. The *hormone* produced in the pancreas. It controls *sugar metabolism* in the body, on being injected, it lowers the *blood* sugar content and relieves the symptoms of diabetes mellitus. Insulin in one of the few *proteins* whose detailed structures is known.

Intelligence. Mental ability, power of knowing, learning.

Interface. Surface which separates two chemical phases.

Interferometer. It divides beam of light into number of beams and re-units these to produce interference; used for testing lenses and prisms.

Intergalactic Space. *Space* between *galaxies,* in which intergalactic *matter* occurs.

Intermediate (Chem.). A compound used in an intermediate step in the manufacture of a final product by chemical *synthesis.*

Intermediate Frequency. In *superheterodyne radio* receivers, the *carrier wave frequency* of the incoming radio wave is changed to a fixed intermediate frequency by *heterodyne* action, for ease of amplification before detection.

Intermediate Neutrons. *Neutrons* having *kinetic energies* between those of *epithermal* and *fast neutrons, i.e.,* between 100 *eV* and 0.1 *MeV.*

Intermediate Vector Boson. W-particle. The *virtual particle* that has been postulated as the particle exchanged during *weak interactions.*

Intermetallic Compound. A compound in which two or more metals are held together by metallic bonds.

Internal-combustion Engine. An *engine* in which *energy* supplied by *fuel* is directly transformed into *mechanical energy* by the controlled *combustion* of the fuel in an enclosed cylinder behind a piston, *e.g., Otto engine,* the *Diesel engine.*

Internal Conversion. A process in which an *exicted nucleus* decays to the *ground state,* the energy released is used to eject a *conversion electron* from an inner shell of the atom. The excited *ion* so formed **emits** *photon (X-ray)* or an Auger electron.

Internal Energy. Thermodynamic energy. E. The total *energy* associated with a system, which cannot itself usually be determined. However, the change in the internal energy of a system, ΔE is a useful *thermodynamic* quantity, and is defined by $\Delta E = Q—W$ where Q is the *heat* absorbed by the system from its surroundings and W is the *work* done on the surroundings.

Internal Resistance. The resistance inside an *electric cell* or other source of *current.* It is equal to $E\text{-}V/I$.

Internal Standard Line. In *spectrographic analysis* an internal standard line is a line within the *line spectrum* of the material being analysed, due to a known amount of an *element* present in, or added to, the material.

Internal Stress. The *stress* within a solid material *e.g., metal, glass* etc., due to heat treatment, cold working, or non-uniform molecular structure.

International Candle. The former unit of *luminous intensity*. It has now been replaced by the *candela.*

International Date Line. An imaginary line on the surface of the *Earth* joining the North and South poles, approximately following the 180° *meridian* through the Pacific Ocean. This line is used to mark the internationally agreed start of a calendar day. Crossing from east to west a traveller changes the day to the next day, and crossing it from west to east goes one day back.

International Practical Temperature Scale. It consists of a practical scale of *temperature* defined so that it conforms as closely as possible to the *thermodynamic temperature*. The unit of temperature is the *kelvin* (symbol K.)

Inulin. Indigestible starch; used in studies of kidney function, it is freely filtered and not actively transported.

Invar. An *alloy* containing 63.8% iron, 36% nickel, 0.2% carbon. It has a very low coefficient of *expansion*. Used for balance wheels of watches and other accurate instruments.

Inverse Square Law. The intensity of an effect at a point B due to a source at A varies inversely as the *square* of the distance AB.

Inverse Trigonometrical Functions. If $y=\sin x$, then the inverse-trigonometrical function of x is $\sin^{-1} y$ where $\sin^{-1} y$ is the *angle* whose sine is y.

Inverse Variation. One quantity is said to be inversely proportional to another, if the *product* of the two is a constant.

Inversion. Change from one optical isomer to other.

Invertase. Enzyme contained in yeast, it converts cane-sugar into glucose and fructose.

Inverter. Device used to convert DC (direct current) into AC (alternate current).

Invert Sugar. Mixture of fructose and glucose in equal proportions, obtained by inversion of cane-sugar.

In Vitro. In glass or test tube.

In Vivo. In living organisms.

Iodate. Salt of iodic acid.

Iron (Ferrum, Fe). White magnetic metal, exists in three crystalliane forms (alpha, gamma and delta), boiling point 2688°C. It is extracted from its ore by blast furnace process.

Iron (Ferric) Chloride ($FeCl_3$). Brownish-yellow deliquescent crystaline salt obtained on passing stream of dry chlorine on heated iron fillings, used as laboratory reagent and in medicine.

Iron (Feric) Oxide (Fe_2O_3). Reddish-brown insoluble salt that occurs in nature as haematite; used as mordant and pigment.

Iron (Ferric) Sulphate [$Fe_2(SO_4)_3$]. Yellow hygroscopic compound obtained by heating acidified ferrous sulphate with hydrogen peroxide, used volumetric analysis.

Iron (Ferrous) Chloride ($FeCl_2$). Greenish-white deliquescent compound obtained by heating iron fillings in stream of dry hydrogen chloride, soluble in water and methyl alcohol; used as laboratory reagent.

Iron (Ferrous) Oxide (FeO). Black powder, soluble in dilute acids, formed by reduction of ferric oxide with the help of carbon monoxide or hydrogen, converts into ferric oxide on being exposed to air.

Iron (Ferrous) Sulphate ($FeSO_4 . 7H_2O$) Green vitriol; obtained by dissolving scrap iron in dilute sulphuric acid; used in dyeing and tanning and as laboratory reagent in volumetric analysis.

Iron (Ferrous) Sulphide (FeS). Precipitate obtained on passing hydrogen sulphate, through alkaline solution of ferrous salt; commerically prepared by heating iron fillings with sulphur $Fe+S \rightarrow FeS$; used for preparation of H_2S gas in laboratories.

Irradiance E. The *radiant flux* striking unit area of surface.

Irradiation. Exposure to *radiation* of any kind. Artificial *radioisotopes* are made by irradiation of stable isotopes with *neutrons* in a *nuclear reactor*. Even small does may be used for sterilization of food due to the sensitivity of biological *cells* to irradiation by *ionizing radiation.*

Irreversible Process. Any process, exept one that is a completely *reversible process.*

Irreversible Reaction. A *chemical reaction* that proceeds to compe-tion; the resulting products do not react to form the original substances.

Irritability. The property of living *organisms* which enables them to respond to external stimuli.

Isatin ($C_8H_5O_2N$). Red solid, made by oxidising indigo with nitric acid.

Isinglass. Material prepared from gelatin (about 90%), obtained from sturgeons' air-bladder; used for clarifying alcoholic beverages.

Isobars. Nuclides having same mass number but different atomic numbers.

Isobaric Surface. A surface having equal *pressure.*

Isochore. A line that graphically represents the relationship between the *pressure* and the *temperature* of a *liquid* or *gas*, at constant *volume.*

Isocline. A line that joins points of equal angle of *magnetic dip.*

Isocyanate. A *compound* containing the NCO group.

Isocyanic Acid HN = C = O. An unstable *isomer* of *cyanic acid*, forms stable *salts* called *isocyanates.*

Isodiapheres. *Nuclides* having the same difference between the number of *neutrons* and *protons.*

Isodimorphism. The phenomenon of a *dimorphous* substance being isomorphous with another dimorphous substance in both its forms.

Isodynamic Line. A line passing through points of equal *horizontal intensity* of the earths magnetic field.

Isoelectric Point. The pH *value* at which a substance or system *e.g., a protein solution* is electrically *neutral*; at this value *electrophoresis* does not occur on applying a direct *electric current.*

Isogonal Line. Line passing through points of equal *magnetic declination.*

Isogonism (chem.). A type of *isomorphism* in which two substances show little or no chemical resemblance but have the same crystalline form.

Isokom. Line joining points of equal *viscosity* on a *phase diagram.*

Isoleucine. A colourless crystalline *amino acid.*

Isomegethic Solutions. *Solutions* composed of *solute molecules* of the same size.

Isomeric. Exhibiting *isomerism.*

Isomerism. 1. Two or more chemical *compounds* having the same *molecular formula* but having different properties owing to a different arrangement of *atoms* within the *molecule, e.g.,* ammonium cyanate, NH_4CNO, and urea $CO(NH_2)_2$ are isomers. 2. In *nuclear physics, nuclei* having the same *atomic number* and the *mass number*, but existing in different *energy* states, are said to isomeric, *e.g.,* a nucleus in its *ground state* and a nucleus in a *metastable* excited state are isomers.

Isometric. 1. Referring to a system of *crystallization* in which the *axes* are at right angles to each other. **2.** A method of projecting a drawing (isometric projection) in which the three axes are equally inclined to the surface of the drawing, and all lines are drawn to scale.

Isomorphism. Similarity of crystalline form, generally indicating similar or analogous chemical composition; *e.g., alums* are isomorphous.

Isooctane. 2, 2, 4-trimethylpentane. $(CH_3)_3CCH_2CH(CH_3)_3$. The *isomer* of *octane* used in defining *octane numbers.*

Isophthalic Acid. Benzene- 1 2- dicarboxylic acid $C_6H_4(COOH)_2$. The *meta-isomer* of *phthalic acid,* m.p. 345– 7°C., used in manufacture of synthetic *resins* and *plasticizers.*

Isoprene. 2-methylbuta-1, 3-diene. $CH_2CH.C(CH_3):CH_2$. A colourless *liquid,* b.p. 34°C. Natural *rubber* consists mainly of a polymer of isoprene and it is used in making synthetic rubbers.

Isosceles Triangle. A *triangle* having two equal sides .

Isotherism. The phenomenon of substances having *molecules* with the same number of *atoms* and the same total number of *electrons;* leading in similarity in physical properties. *e.g., carbon dioxide,* CO_2 and *nitrous oxide,* N_2O.

Isotherm. 1. Isothermal line. A line connecting points at an equal *temperature.* **2.** A relationship (graphical or mathematical) between two *variables* at constant temperature.

Isothermal Change. Isothermal process. A change on process taking place at constant *temperature. e.g.,* the isothermal *expansion of gas.*

Isotones. *Nuclides* containing the same number of *neutrons* but containing a different number of *protons.*

Isotonic Solutions. *Solutions* having the same *osmotic pressure.*

Isotopes. *Atoms* of the same *element* but differ in *mass number*: The isotopes of an element are identical in chemical properties, and in all physical properties except those determined by the *mass* of the atom. The different isotopes of an element contain different numbers of *neutrons* in their *nuclei.* Almost all elements found in nature are mixture of several isotopes.

Isotopes, Separation of. As the *isotopes* of an *element* have indentical chemical properties but slightly different physical properties, their separation depends upon physical operations. The following methods are used: *diffusion* (either *gaseous* or *thermal*); *distillation*; centrifuging of *gases* or *liquids; electrolysis.*

Isotopic Number. Neutron excess. The difference between the number of *neutrons* and the number of *protons.*

Isotopic Spin Isospin. Isobaric spin. The word 'spin' is not intended to imply any conventional image of rotation in this context, it is used in analogy to *angular momentum* to which the concetp of isotopic spin bears a close resemblance. Isotopic spin is conserved in all *strong nuclear interactions.*

Isotopic Mass. The *relative atomic mass* of an individual *isotope.*

Isotropic. Exhibiting uniform properties throughout.

- Ite. A suffix denoting a *salt* of the corresponding ous *acid e.g., sulphite* from sulphurous acid.

Ivory Black. A form of carbon obtained from *animal charcoal.*

J

Jasper. Precious stone. Coloured impure form of natural silica (SiO_2).

Javelle Water. Eau de Jevelle, solution containing potassium hypochlorite (KOCl); obtained by action of chlorine on cold potassium hydroxide (KOH); used for bleaching and as disinfectant.

Jet Engine. Gas turbine which produced stream of hot gas enabling aircraft or boat to be propelled through air or water.

Josephson Effects. Effects that occur when two superconductors are separated by a thin insulated layer. At normal *temperatures* a small current can flow between the *conductors* as a result of the *tunnel effect.* The Josephson junction has proved useful as research tool and also as a *logic* component and special purpose switch.

John Reductor. Tube containing zinc amalgam; used for reducing solutions prior to titration or estimation.

Joint. The point in a skeleton where two or more bones meet and movement may be possible.

Joule (J). SI unit of works or energy–one joule=one newton metre or one watt second or 10^7 ergs or 0.238846 calorie.

Joule Effect. Heating effect produced due to flow of current through resistance.

Joule-Kelvin Effect (Joule-Thomson effect). Change in temperature arising as a result of gas expanding by flooding through porous plug from region of high pressure into region of low pressure.

Joule's Law. At low pressure internal energy of a gas at constant temperature is independent of its volume.

Joule-Thomson Effect. Joule-Kelvin effect. When a *gas* is allowed to expand through a porous plug, a change of *temperature* occurs, proportional to the *pressure* differences across the plug. The effect can give rise to either cooling or heating, depending upon the initial temperature and pressure difference. Gases expanding through a porus plug below their inversion temperature are cooled, otherwise they are heated.

Junction Detector. A detector of *ionizing radiation*. It makes used of a *semiconductor junction*. It produces a current *pulse*, which is proportional to the energy falling on the *depletion region* of the junction, when it is reverse biased. They usually consist of gold-silicon devices and are used in medicine and space research.

Junction Rectifier. A *rectifier* based upon a *semiconductor junction*.

Junction Transistor. A *transistor* having a *base electrode* and two or more electrodes connected to *semiconductor junctions*.

Jupiter (astr.). A *planet*. It has sixteen small *satellities*, with its *orbit* between those of *Mars* and *Saturn*. It is the largest of the planets.

K

K. A symbol used in computing for bilo or sometimes for bilobyte or bilobit.

Kainite ($MgSO_4$. KCl. $3H_2O$). Naturally occurring double salt of potassium and magnesium, contains about 30% potassium chloride.

Kairomone. Chemical emitted by plant or animal that induces response in individual of another species and is beneficial to receiving species e.g., plant smell that attracts insects.

Kaleidoscope. Optical toy, shows ever-changing variety of symmetrical beautiful colours, forms and patterns used by weavers.

Kallikeins. Proteolytic enzymes that help in formation of kinins from kininogens.

Kaolin. Clay mineral — anauxite, dickite, kaolinite, nacrite etc.

Karyo. Denoting nucleus of a cell or its contents.

Karyogamy. Fusion of two nuclei, each derived from two parents during sexual reproduction that results in formation of zgyote nucleus.

Katharometer. Device used for measuring thermal conductivity, particularly as a detector in gas chromatography.

K-band. Band of radio frequencies that extend from 10900 to 36000 megahertz.

Keepers of Magnets. Short iron bars used to prevent permanent *magnets* from losing their *magnetism.*

Kekule Formula. The graphic representation of *benzene* first suggested by F.A. Kekule von Stradonitz (1829-96).

Kelp. Sea-weed or its *ash*. Used as a source of iodine.

Kelvin. The *SI unit of thermodynamic temperature* defined as the fractions 1/273.16 of thermodynamic temperature of the *triple point of water*, i.e. the triple point of water contains exactly 273.16 kelvins.

Kelvin Temperature. *Temperature* expressed in *kelvins*. The same as the thermodynamic temperature. Symbol K.

Kepler's Laws. **1.** The *planets* move about the *Sun* in *ellipses*, at one focus of which the Sun is situated. **2.** The *radius vector* joining each planet with the Sun describes equal areas in equal times. **3.** The ratio of the *square* of the planet's year to the cube of the planet's mean distance from the Sun is the same for all planets.

Keratin. A *protein* that is the principal constituent of wool, hair, horns, and hoofs.

Kernite ($Na_2B_4O_7.H_2O$). Mineral ore of boron.

Kerosene. Paraffin hydrocarbon obtained by distillation of crude petroleum; used as liquid fuel.

Ketal. Organic compound obtained by addition of alcohol to ketone.

Ketene ($CH_2C : CO$). Gas obtained by decomposition of acetone or acetic anhydride.

Ketone. Organic compound containing carbonyl group attached to two like or unlike alkyl radicals.

Ketose. Monosaccharide containing a ketone group.

Ketosis. Accumulation of ketone bodies in blood streams due to ability or tissues to oxidise ketones is exceeded on account of metabolic disorder.

Ketosterone. Intermediate compound formed during biosynthesis of estriol from estrone.

Killikrein. Proteolytic enzyme that converts, kininogens to kinins.

Kilo. Thousand, 10^3.

Kilocycle. Measure of electromagnetic wave frequency (=1000 cycles per second).

Kiloton Bomb. Nuclear bomb having explosive power equivalent to 4×10^{12} joules.

Kilovolt. Unit of electric potential (=1000 volts.).

Kinase. Enzyme, activates inactive form of other enzymes.

Kindling Temperature. Minimum temperature at which a material catches fire and burns.

Kinescope. Cathode-ray or picture tube used in television receiver.

Kinetic Energy. Energy possessed by the body due to its motion.

Kinetics. Study of rates of chemical reactions.

Kinetic Theory of Gases. A mathematical explanation of the behaviour of *gases* on the assumption that gases consist of *molecules* in ceaseless motion in space, the *kinetic energy* of the molecules depending upon the *temperature* of the gas; the molecules of a *perfect gas* are considered to be elastic particles that collide with each other and with the walls of the containing vessel.

Kink Instability. In a *thermonuclear reaction* an instability in the magnetically confined *plasma* resulting from a local deformation of the plasma. The kink has a tendency to grow because the *magnetic lines of forces* of the self-induced confining field are crowded on the *concave* side of the kink.

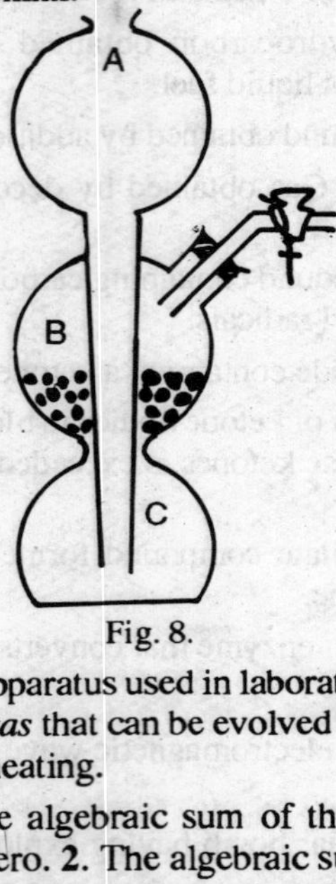

Fig. 8.

Kipp's Apparatus. A apparatus used in laboratories for the production of a supply of any *gas* that can be evolved by the action of a *liquid* on a *solid* without heating.

Kirchhoffs Laws. 1. The algebraic sum of the *electric currents* that meet at a point is zero. **2.** The algebraic sum of the *electromotive forces* in any closed circuit is equal to the algebraic sum of the products of the *resistances* of each portion of the circuit and the currents flowing through them.

Kish. A variety of *graphite* sometimes formed in iron *smelting* furnaces.

Kjeldahl Flask. A round - bottomed glass flask with a long wide neck used in the estimation of nitrogen by the *Kjeldahl method.*

Kjeldahl Method. Analytical method of determining the nitrogen content of an *organic* compound.

Klinstat. Apparatus used for studying plant growth in absence of directional stimulus.

Klystron. Electron tube used to generate or amplify electromagnetic radiation in microwave region, by velocity modulation.

Knocking. Process of violent *explosions* in cylinder of a petrol engine, often due to over-compression of the *mixture* of air and petrol vapour ahead of the flame front.

Knock-on Collission (phys.). A process in which an *elementary particle* or *nucleus* is set in motion on being struck by another high-*energy* particle (or *proton*). Also used in relation to collision due to which an *electron* is knocked out of its atomic *orbit* by some other particle.

Knot. A unit speed of equal to 1 *nautical mile* per hour. (Approximately 1.15 statute miles per hour).

Kohlrausch's Law. At infinite dilution the *conductivity* of an *electrolyte* is equl to the sum of the conductivities of the *ions* into which the substance dissociates.

Kovar. An *alloy* of cobalt, iron and nickel, having *expansivity* similar to that of glass. It is used for glass-to metal seals in *thermionic valves* and *transistors.*

Kreb's Cycle. Complex cycle of enzymatic reactions involving oxidation of two carbon acetyl units to carbon dioxide and water.

Kroll Process. It is reduction of metal halides by magnesium, sodium calcium.

Kryptol. Mixture of carborundum, clay and graphite ; used as electrical resistance in electric furnaces.

Kryptron (Kr). Colourless, odourless, monoatomic gas, boiling point 164°C, used for electric bulbs and fluorescent lights.

Kundt Effect. Rotation of plane of vibration of polarised light on traversing an isotropic transparent medium placed in magnetic field possessing a component in direction of light ray.

Kundt's Tube. Device for measuring speed of sound.

Kupfer-nickel. Nickle arsenide (NiAs); important ore of nickle.

Kurrol Salt ($NAPO_3$). Crystalline high temperature form of sodium metaphosphate, obtained by annealing Graham's salt at 55°C.

Kymograph. Apparatus used for recording graphically various physiological moments in animals—blood—pressure, heart-beat, movements of lungs etc.

L

Labelled Compound. Compound in which a stable atom is replaced by radioactive isotope of that very atom.

Labile. Complex that exchanges ligands at a rapid rate.

Laboratory (Lab). Room for scientific experimental work or research.

Lachrimal. Relating to tears.

Lachrymal. A Duct that conveys tear-water from inner corner of eye to nose.

Lachrymal Glands. Glands situated near outer angle of eyes of vertebrates that secret tears.

Lachrymator. Any substance that causes tears to flow *e.g.*, tear-gas.

Lachteal. Pertains to milk .

Lacquer. Varnish of lac and alcohol.

Lactase. Enzyme present in intestinal juices, it catalyses production of glucose and galactose from lactose.

Lactation. Secretion of milk by mammary glands; period of suckling.

Lactic Acid [CH_3CH (OH) CO_2H]. An acid obtained from milk (lactose) by bacterial action.

Lactiferous. Conveying or producing milk or milky juice.

Lactifluous. Flowing with milk.

Lactogenic. Hormone that stimulates milk production.

Lactometer. Instrument used for measuring purity of milk.

Lactose ($C_{12}H_{22}O_{11}$). Milk-sugar; hard gritty, crystalline, soluble disaccharide, present in milk of all mammals.

Laevorotatory. Rotating the plane of vibration of *polarized light* to the left.

Laevulose. Fructose, fruit sugar. $C_6H_{12}O_6$.

Lagena. That portion of membranous labyrinth which is concerned with hearing.

Lake. Used dyeing, refes to a coloured insoluble substance formed by the chemical combination of a soluble dye and the mordant.

Lamarckism. The theory put ward by Lamark which states that acquired characters are inherited during sexual reproduction.

Lambda Particle. An *elementary particle*, a *hyperon*, that has no charge and is 2183 times heavier than an *electron.*

Lambda point λ. The *temperature* below which helium becomes *superfluid.* λ=2.186 K.

Lambert. A former unit of *luminance.* 1 lambert = 3180 candela per square metre.

Lambert's Law of Illumination. The *iluminance* of a surface upon which the *light* falls normally from a *point source* is inversely proportional to the *square* of the distance between the surface and the source.

Lamb Shift. A small difference in the *energy levels* $^2S_{1/2}$ and $^2P_{1/2}$ of the *hydrogen spectrum* resulting from the quantization of the interaction between the atomic *electron* and the *electromagnetic field.*

Lamina. A thin sheet.

Laminar Flow. The flow of *fluid* following closely the shape of a stream lined without turbulence.

Laminarin. It is a polysaccharide storage–product and is found in large brown algae.

Laminated Iron. Thin sheets of iron used for cores of *transformers.* The sheets are varnished to give a high resistance between them, in order to reduce losses due to *eddy currents.*

Lamp-Black. Syot: *allotropic* form of powdered *carbon.* It is used as a *pigment* and *filler.*

Lanolin. A *wax*-like material obtained from wool-grease and containing *cholestrol*, $C_{27}H_{45}OH$, and other complex *organic* substances, readily absorbed by the skin used in ointments and cosmetics.

Lanthanides. Lanthanons, lanthanoids rare earths. A group of rare metallic *elements* with *atomic numbers* from 57 to 71 inclusive.

Lanthanum La. Element. R. a.m. 138.91. At No. 57. A silvery metal, r.d. 6.2, m.p. 920°C., b.p. 3464°C., used in *pyrophoric alloys* and as a *catalyst* in oil cracking.

Lapis Lazuli. Sodium aluminium silicate containing sulpnur. A rare mineral having beautiful blue colour.

Laplace Operator. Laplacian ∇_2. The *differential* operator that gives the sum of the *partial drivatives* of second *order* with respect to each *variable i.e.*

$$\nabla^2 u=\delta^2 u/\delta x^2+\delta^2 u/\delta y^2+\delta^2 u/\delta z^2 = 0.$$

This *equation* is known as the 'Laplace equation'.

Large Calorie. Kilogram - calorie, Calorie. = 1000 *calories.*

Larmor Precession. The orbital motion of the *electron* about the *nucleus* of an *atom* usually gives the atom a *resultant angular momentum* and a *magnetic moment.* These two properties cause the atom to precess (see *precessional motion*) about the direction of any applied *magnetic field.* This is Larmor precession.

Laser. Light amplification achieved by stimulated emission of radiation. It is a device capable of producing powerful monochromatic and coherent beam of light; used in medicine, scientific warfare etc.

Latex (pl. latices). Milky liquid found in some flowering plants contained in especial cells (laticifers); consists of many substances in solution and colloidal suspension; used in medicines and also in rubber and other industries.

Latin Square. A square array of numbers or letters in which each number or letters appears once in every row or column. For example,

1	2	3	4		p	q	r	s
4	3	2	1	but not	q	r	s	p
3	1	4	2		r	q	p	s
2	4	1	3		s	p	r	p

Lattice Energy. Amount of energy released during formation of one mole of crystal of solid from constituent gaseous irons separated from each other by infinite distance.

Laudanum. Tincture of opium, reddish-brown liquid; It is a pain reliever.

Laughing Gas (N_2O). Dinitrogen oxide or nitrous oxide, colourless gas used as mild anaesthetic in dentistry.

Lawrencium (LW). Radioactive transuranic element of actinide series obtained by bombardment of californium with accelerated ^{10}B ions.

Leaching. To allow liquid to percolate through; washing out a soluble constituent.

Lead (Pb). Shoft silvery -white metal having bluish tinge, obtained by reduction of roasted ore with coke, boilling point 1755°C, reacts slowly with most strong acids; used in manufacture of batteries, cables, pigments etc.

Lead Chloride ($PbCl_2$). White precipitate obtained on adding lead ion to hydrochloric acid or some soluble chloride.

Lead Chromate ($PbCrO_4$). Yellow precipitate obtained by mixing solution of lead salts and potassium chromate: used as yellow pigment.

Lead Dioxide (PbO_2). Chocolate-brown powder obtained by fusing mixture of sodium nitrate and sodium chlorate or by passing chlorine into hot aqueous suspension of lead sulphide and magnesium hydroxide.

Lead Iodide (PbI_2). Fine yellow crystalline substance obtained by dissolving lead oxide or lead carbonate in hydro-iodic acid.

Lead Monoxide (PbO). Yellowish brown powder obtained on heating lead carbonate, hydroxide or nitrate; used in preparing oil paints, varnishes etc.

Lead Sulphate ($PbSO_4$). White solid, found in nature as anglesite, can be prepared by adding dilute sulphuric acid to solution of lead salt; used in gravimetric estimation of lead.

Lead Sulphide (PbS). Black precipitate obtained on passing hydrogen sulphide gas in solution of plumbous salt; occurs in nature as galena.

Lead Tetra-ethyl. Extremely poisonous, colourless liquid, obtained by heating ethlychloride with sodium-lead alloy under pressure; It is insoluble in water but soluble in ether. Used as anti-knock agent to improve quality of petrol.

Leblanc Process. Salt-cake process. Old process for the manufacture of *sodium carbonate*, Na_2CO_3.

Le Chatelier Principle. If a system in *equilibrium* is subjected to a change the system tends to react in such a way as to oppose the effect of the change.

Lecithins. Naturally occurring complex *lipids* consisting of *glycerides* in which one of the *acyl* groups is replaced by a *phosphorylcholine group*; they are similar to *fats*, but contain nitrogen and phosphorus as well.

Leclanche Cell. A *primary* cell with a positive *electrode* surrounded by a mixture of *manganese dioxide* and powdered carbon in a porous pot. This stands in a *solution* of *ammonium chloride*, in a jar, which also contains the negative electrode of zinc. The *E.M.F.* is approximately 1.5 *volts*. Leclanche cells are widely used for many purposes which require an intermittent current. The common *dry cell* is a special form of Leclanche cell.

Lens. Piece of transparent material bound by two surfaces, one of which is curved; used in cameras, spectacles, microscopes, telescopes and other optical instruments.

Lenticular. Relating to a *lens*, especially a *bi-convex* lens, or resembling such a lens in shape.

Lenz's Law. When there is a change in the *magnetic flux* linked with a circuit, the *electric current* induced in the circuit will have a magnetic field opposing the change producing it.

Leprosy. Chronic skin disease caused by a bacillus and occurring in two forms–tubercular, beginning with spots and thickenings of skin, and anaesthetic, attacking nerves with loss of sensation in affected areas of skin.

Lepton. Collective name for electrons, muons and neutrinos.

Lesion. Structural change in tissue due to disease, injury and wound.

Lethal Gene. Mutant allele that kills the bearer.

Leucine. Natural amino acid; with unsubstituted chains, its isomer is isoleucine another essential amino acid.

Leucocidin. Toxic substance released by certain bacteria, it destroys leucocytes.

Leucoma. Refers to white opacity of cornea of eye.

Leucoplast. Starch-forming body in protoplasm.

Leukaemia. Abnormal increase of white blood corpuscles in blood, from about 7000 to 700,000 or more. It is a type of blood cancer.

Lever. A rigid bar capable of free rotation about a fixed point of suppport, the fulcrum.

Lewis Acids and Bases. According to this concept an acid is defined as a substance that can accept *lone pair of electrons*. A base is defined as a substance that can donate a lone pair of electrons.

Lewisite. 1. β-Chlorovinyldichloroarsine. $ClCH{:}CHAsCl_2$. An oily liquid, b.p. 190°C. **2.** The mineral calcium titanium antimonate, 5 $CaO.2TiO_2, 3Sb_2O_5$.

Leyden Jar. A form of electrostatic *capacitor* of historical interest. It was invented in 1745 in the Dutca town of Leyden.

Libration. An oscillation of the Moon's face from side to side. Due to libration about 58% of the Moon's surface can be seen from the Earth.

Life-belt. Buoyant belt to sustain someone in water.

Life-buoy. Buoy to support somebody in water till he can be rescued.

Life-cycle. Functional and morphological stages through which an organism must pass between two successive primary stages.

Lifeless. Inanimate; dead.

Life-span. Longevity; age at death of an animal or plant.

Ligand. Molecule or ion capable of giving a pair of electrons so as to form coordinate bond.

Ligases. Enzymes that catalyse many biological substances.

Light. Illumination; brightness. Form of electromagnetic radiation that acts upon retina of eye thus making sight possible.

Lightning. Flash of light in sky caused by discharge of atmospheric electricity from one cloud to another or between cloud and the earth.

Lightning Conductor. A *conductor* of *electricity* connected to earth ending in one or more sharp points attached to a high part of a building. It provides a direct path of low resistance to earth.

Light Pen. *Computer* device used with a *visual display unit*. When pointed at a *cathode-ray* tube it can sense whether or not the shot is illuminated.

Light-sensitive. Possessing photoconductive, photoemissive or photovoltaic properties.

Light Year. Distances travelled by light in one year (about 100,000, 000,000,000 kilometres).

Lignin. Complex organic substances occurring in woody tissues of plants often combined with cellulose.

Lignite (brown coal). Brownwish-black natural deposit, of recent origin, resembling coal but has higher percentage of hydrocarbon than ordinary coal.

Ligroin. A mixture consisting of various hydrocarbons. It is similar to benzine, boiling range (80-130°C).

Lime. Any chemical and physical form of quicklime, hydrated-lime hydraulic lime; calcium oxide (CaO).

Lime, Air-slaked. Lime that has absorbed carbon dioxide (CO_2) from air; it consists of calcium carbonate ($CaCO_3$) and calcium hydroxide [$Ca(OH)_2$].

Lime Chlorinated [CaCl(OCl)]. White powder obtained by treating slaked lime with chlorine ; used as bleaching powder, deodoriser, disinfectant etc.

Lime, Hydrated. Calcium hydroxide [$Ca(OH)_2$], obtained by treating quicklime with water.

Lime, Hydraulic. Calcined limestone that when pulverised absorbs water without swelling or heating and gives cement.

Limestone. Sendimentary rock of calcium carbonates ($CaCO_3$).

Lime-wash. Milky mixture of slaked lime and water; used for coating walls.

Lime Water. Solution of calcium hydroxide [$Ca(OH)_2$] in water.

Limit of Special Series. The lines appearing in *line spectrum* of any *element* can be grouped into definite series. The shortest *wavelength* of any such series is called the limit of the series.

Limonene. Dipentene $C_{10}H_{16}$. An *optically active* liquid *terpene*, b.p. 176—178°C., it occurs in some *essential oil* used as a *solvent* and in the manufacture of *resins* and *surface-active agents.*

Limonite. A natural *hydrated* from of *iron (III) oxide*, Fe_2O_3. An iron ore.

Linalool $C_{10}H_{17}OH$. A colourless *liquid terpene alcohol* b.p. 198—200°C., occurring in certain *essential oils* used in perfumes.

Linalyl Ethanoate Bergamol. $C_{10}H_{17}COOCH_3$. A colourless *liquid,* b.p. 220°C., having a pleasant odour; used in perfumes and *soaps.*

Lindane. White, crystalline substance containing not less than 99% gamma isomers of BHC (HCH); contact and stomach poison with some fumigant properties; used as insecticide, effective against soil borne insects.

Linde Process. A process for producing *liquid air*, based on the *Joule-Thomson effect.* The process is also used to liquefy other gases, such as hydrogen and helium.

Linear. 1. Arranged in a line. **2.** Having only one dimension. **3.** Heaving only first *degree* terms. **4.** Heaving an output directly proportional to the input.

Linear Absorption coefficient *a*. A measure of a medium's ability to absorb radiation, but not to scatter or diffuse it is given by $\phi x/\phi_o = e^{-ax}$, where ϕ_o is the initial *radiant flux* or *luminous flux* and ϕx is the flux after it has travelled a distance x through the medium.

Linear Acceleator Linac. Apparatus for accelerating *ions* to high *energies.* Consisting of a row of cylindrical *electrodes* separated by small gaps and having a common axis. Alternate electrodes are connected to each other and ***high-frequency potential*** is applied between the two sets of electrodes. The frequency, and the lengths of the different electrodes, are such that the ions are accelerated each time they cross a gap between two electrodes.

Linear Attenuation Coefficient μ. A measure of a medium's ability both to absorb and to diffuse radiation. If a ***luminous flux*** or radiant flux, ϕ_o, passes perpendicularly through a section of the attenuating medium, *x*, being reduced to ϕ_x, then the linear attenuation coefficient, μ is given by $\phi_x/\phi_o = e.^{\mu x -}$

Linear Motor. A form of ***induction motor***, having linear ***stator*** and ***rotor*** instead of cylindrical, and parallel instead of ***coaxial***.

Linear Relationship. A relationship existing between two ***variable*** quantities such that a ***graph*** representing the manner in which they vary with each other will be a straight line.

Line-of-sight Velocity. The ***velocity*** at which a heavenly body approaches, or recedes from, the Earth.

Line Pair. In ***spectrographic analysis*** it consist of the particular spectral line used in the determination of the ***concentration*** of an ***element*** and the ***internal standard line*** with which it is compared.

Line Printer. Device from a ***computer***, it prints a line of ***characters*** at a rate of between 300 and 3000 lines per minute.

Line Spectrum. A ***spectrum*** (***emission*** or ***absorption***) consisting of definite single lines, each corresponding to a particular ***wavelength***; it is characteristic of an ***element*** in the atomic state.

Linkage. Association of two or more non-allelomorphic genes, that pass from generation to generation as an inseparable unit and do not exhibit any independent assortment being in same chromosome.

Linoleic Acid ($C_{17}H_{31}COOH$). A yellow oily liquid ***unsaturated fatty acid***, b.p. 229°C., occurs in ***vegetable oils, linseed oil.***

Linseed Oil. A ***vegetable oil*** extracted from the seeds of flux plants. Contains ***glycerides*** of ***oleic acid*** and other ***unsaturated fatty acids.*** Widely used in the ***paint*** and varnish industries and for manufacturing linoleum.

Lipase. An ***enzyme*** that hydralises (see ***hydrolysis***) ***fats.***

Lipids, Lipoids. A group of ***organic compounds insoluble*** in ***water soluble*** in many organic ***solvents.*** They are usually divided into three groups : (1) 'Simple lipids', ***fats and oils*** as well ***waxes*** ; (2) 'Compound lipids', such as ***phospholipids and glycolipids*** ; (3) 'Derived lipids', such as the ***steroids.***

Lipochrome. Pigment of butter, fat etc.

Lipoclastic. Lipolytic, Enzymes that can hydrolyse fats into fatty acids and glycerin.

Lipocyte. Fat-bearing cell.

Lipogenesis. Formation of fat from non-fatty sources in one's body.

Lipoma. Fatty tumour.

Lipomatosis. Refers to excessive growth of fat.

Lipoproteins. Cholesterol and other fatty substances which are associated with certain water soluble proteins in the body particularly in brain and nerve tissue.

Lipowitz Alloy. Fusible alloy, melting point 70°C, contains bismuth (50%), Lead (27%), tin (13%) and cadmium (10%).

Liquation. Method of removing impurity which is less fusible than the metal. Impure metal is placed on sloping hearth of a furnace and gently heated; metal melts and drains away, leaving behind the infusible material on hearth.

Liquefication. Change in state of substance to liquid.

Liquefied Gas. Gaseous substance converted to liquid by compression or cooling *e.g.*, liquefied petroleum gas.

Liqueur. Alcoholic preparation flavoured and sweetened *e.g.*, brandy, sherry etc.

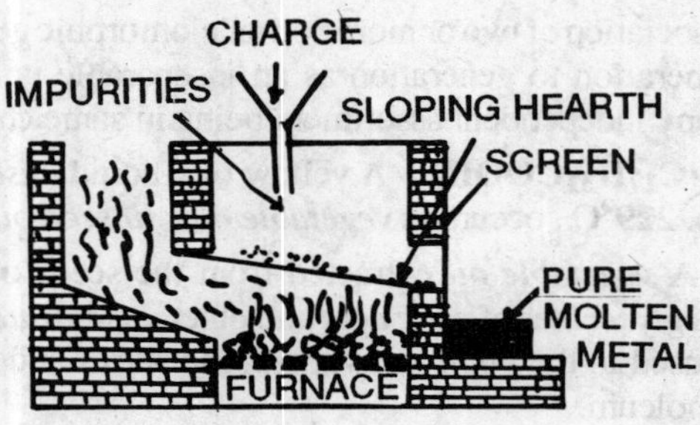

Liquation

Liquid. Fluid; physical state of matter between solid and gas, liquid molecules move freely about one another but do not fly apart.

Liquid Air. Pale blue air containing liquid oxygen, boiling point—183°C, and liquid nitrogen boiling point—166°C.

Liquid-Crystal Display. A *digital display* in an electronic calculator, etc., based on *liquid-crystal* cells that change their reflectivity in an applied *electric field.*

Liquid Crystals. Relatively charge regions of regularly arranged *molecules* in *liquds* which are analogous to crystals and sufficiently distinct from the bulk liquid to constitute identifiable 'meso-phases'.

Liquid Drop Model of the Nucleus. A hypothetical model of the atomic *nucleus* properties of nucleus are compared to those of a drop of *liquid.*

Lissajous Figure. The *locus* of the *resultant* displacement of a point on which two or more simple periodic motions are impressed.

Litharge. *lead oxide* PbO.

Lithium. **Li.** Element. R a.m. 6.939. At. No 3. A light *alkali metal,* m.p. 179°C., b.p. 1340°C., r.d. 0.534. It is the lightest *solid* known. It is used in *alloys.*

Lithium Aluminiumhydride ($LiAlH_4$). White solid prepared by action of anhydrous aluminium chloride on lithium hydride in ethereal solution; a powerful reducing agent.

Lithium Carbonate (Li_2CO_3). White precipitate formed when ammonium carbonate is added to a solution of soluble lithium salt; used in glass industry.

Lithium Chloride (LiCl). White solid obtained by dissolving lithium carbonate in dilute hydrochloric acid and crystallising the product; used in air-conditioning plant to regulate humidity of air.

Lithium Fluoride (LiF). Sparingly soluble salt prepared by action of hydrofluoric acid on lithium carbonate; used for pottery glazes, acid resisting coating etc.

Lithium Hydrogen Carbonate ($LiHCO_3$). Known only in solution; produced when carbon dioxide is passed through suspension of lithium carbonate; used in medicines for treatment of gout.

Lithium Nitrate ($LiNO_3$). Obtained by dissolving lithium carbonate in nitric acid, on heating decomposes to give lithium oxide (Li_2O); soluble in alcohol and water.

Lithium Nitride (Li_3N). Obtained by heating lithium in current of nitrogen; reacts with water to yield lithium hydroxide and ammonia, $Li_3N + 3H_2O \rightarrow LiOH + NH_3$.

Lithium Oxide (Li_2O). White powder obtained by heating lithium in air, $4\ Li + O_2 \rightarrow 2Li_2O$.

Lithium Phosphate (Li_3PO_4). Precipitate obtained when soluble phosphate is added to solution of lithium salt.

Lithium Sulphate (Li_2SO_4). White solid obtained by addition of excess of lithium oxide or lithium carbonate to sulphuric acid; highly soluble in water.

Lithopone. Mixture of zinc sulphide (ZnS) and barium sulphate ($BsSO_4$); used in paints as non-poisonous substitute for white lead.

Lithotroph. Organism that manufactures food by using energy of oxidation of inorganic raw materials.

Litmus. Soluble purple substance of plant orgin; turns red in acid medium and blue in alkaline medium.

Litre. Unit of volume in metric system (=1000 cubic centimetres of millilitres of water).

Liver of Sulphur. A *mixture* of *sulphides* and other sulphur *compounds* of potassium, obtained by fusing *potassium carbonate,* K_2CO_3, with sulphur. It is used as an *insecticide* and *fungicide* in gardening.

Living Fossil. Present day species that resembles in some characteristics to extinct organisms only.

Lixiviation. Treating mixtures of soluble and insoluble minerals with solvents so as to dissolve the same partially; used for extracting precious metals from their ores.

LNG (liquid natural gas). Natural gas liquefied at atmospheric pressure by cooling to about—160°C; used for transportation purpose.

Lodestone. A magnetic variety of natural *iron oxide.* Fe_3O_4, *magnetite.*

Logarithmic Scale. A scale of measurement in which an increase of one unit represents a tenfold increase in the quantity measured.

Logarithms. If a number, *a*, is expressed as a *power* of another number, *b, i.e.,* if $a=b^n$, then *n* is said to be the logarithm of *a* to *base b*, written $\log_b a$. Common logarithms are to base 10. Natural or Napierian logarithms are to the base 'e' (which has the value 2.71828). $\log_e a=2.303 \log_{10} a$.

Logic. In digital *computers* the logic circuits are the basic switching circuits now formed in *microchips* and *integrated circuits* (formerly they used *transistors* and before that *thermionic valves*).

Lone pair of Electrons. A pair of unshared *electrons* responsible for the formation of coordinate bonds. They occupy the same *orbital* but have opposite *spins.*

longitude. The angle that the terrestrial *meridian* through the geographical poles and a point on the Earth's surface makes with a standard meridian (through Greenwich) is the longitude of the point.

Longitude, Lines of. Imaginary *meridians* on the Earth's surface, they are *great circles* of the Earth interesting at the poles.

Longitudinal Lengthwise; in a line with the length of the object under consideration.

Longitudinal Waves. Waves in which displacement takes place in the direction of propagation of the waves; *e.g., sound* waves, *transverse waves.*

Loran. Radio navigation system for aeroplanes or ships, in this system signals from different fixed stations are received and time intervals compared.

Lorentz-Fitzgerald Contraction. A contraction in the length of a moving object. The contraction is only appreciable at speeds comparable to the *speed of light.*

Lorentz Transformation. A set of *equations* that correlate *space* and *time coordinates* in two *frames of reference*, especially at *relativistic velocities.*

Loschmidt's Constant. The number of *molecules* per unit volume of a *perfect gas* at *S.T.P.*; equal to $2.687\ 19 \times 10^{25}\ m^{-3}$.

Loudness of Sound. The magnitude of the physiological response of the ear to a sound. The loudness of a sound will depend on its frequency. Loudness can be conveniently compared by the units *decibel* and *phon.*

Loudspeaker. A device used for converting *electric currents* into *sounds* loud enough so that it can be heard at a distance.

Lovibond Tintometer. A *colorimeter* for comparing the *colour* of a liquid, surface, powder, or light source with a series of glass slides of standardized colours.

Low Frequency. LF. A *radio frequency* lying in the range 30—300 *kilohertz.*

Lowry-Bronsted Theory. According to this theory *acids* are regarded as *proton* donors and bases as proton acceptors.

LPG (liquid petroleum gas). Mixture of hydrocarbons obtained from petroleum refining operations; used as domestic and industrial fuel. Commercially two mixtures are available—butane (85%) with some propane and pentane; propane (92%) with some butane and ethane.

LSD (lysergic acid diethylamide). Psychomimetic agent and serotonin antagonist; produces transient hallucinations and other mental abberrations.

Luciferase. Oxidising enzyme responsible for light production in certain insects like fire-flies.

Luciferin. Protein-like substance present in luminous organs of certain animals.

Lucifugous. Shunning light.

Lucigen. Lamp burning oil mixed with air in a spray.

Lumen. The *SI unit* of *luminous flux*. The amount of *light* emitted per second in unit *solid angle* of one *steradian* by a uniform *point source* of one *candela* intensity. Symbol 1m.

Luminance. L. The *luminous intensity* of any surface in a given direction per unit of area of that surface, on a plane perpendicular to the given direction.

Luminescence. The emission of *light* from a body except due to high *temperature*. It is caused by the emission of *photons* when an *excited* atom returns to the *ground state*. *Fluorescence* and *phosphorescene* are particular cases of luminescence.

Luminosity. 1. The property of emitting *light*. **2**. The amount of light emitted by a *star*.

Luminous Fus ϕv. It is the amount of *light* passing through that area in one *second*. *SI unit* of luminous flux is the *lumen*.

Luminous Intensity *I*v. The amount of *light* emitted per second in unit *solid angle* by a *point source*, in a given direction. The *SI unit* is the *candela*.

Luminous Paint. *Paint* containing some phosphorescent *compounds* such as *calcium sulphide*, etc. It glows after exposure to *light*.

Lunar Caustic. *Silver nitrate*, $AgNO_3$.

Lunation. Synodic month. Equal to 29 days 12 hours and 44 minutes.

Lute. Cement, clay or any other material used to coat porous surface or a joint so as to make it air-tight and impervious to gas or liquid.

Luteolysis. Regression of corpus luteum so as to begin next menstrual cycle.

Luteotropin. Leuteotropic hormone (LTH), stimulates milk secretion and maternal behaviour.

Lutetium (Lu). Element of rare-earth group, melting point 1652°C.

Lux (lx). Sl unit of illuminance (1 Lx=one lumen per square metre).

Lyddite. Explosive material consists of picric acid, nitro-benzene (10%) and vaseline (3%).

Lyman Series. Series that occurs in the *ultraviolet* region of the *spectrum* of hydrogen.

Lyophilic Colloid. 'Solvent-loving colloid'.

Lyophobic Colloid. 'Solvent-hating colloid'.

Lysergic Acid. $C_{15}H_{15}N_2COOH$. A crystalline substance obtained from ergot, used in the preparation of the hallucinogen LSD.

Lysine. An essential *amino* acid, m.p. 224°C.

Lysis. Process of dissolution or destruction of *cells* (especially *blood cells* or *bacteria*) by *antibodies* called lysins.

Lysol. Soapy solution, contains 50% isomeric cresoles; used as disinfectant.

Lysolecithin. Lipase-A in pancreatic juice it splits fatty acid present in lecithin forming lysolecithin. It damages cell membranes and causes disruption of pancreatic tissue and surrounding fat.

M

Macerate. To soften or break up fibrous substance by dipping in water at room temperature.

Machine. A device in which a certain force applied at one point gives an output force, elsewhere. Simple examples are levers, pulleys, and hydraulic machines.

Machmeter. An instrument used for measuring the speed of an aircraft relative to the speed of *sound.*

Mach number. The ratio of the speed of a *fluid* or body to the local speed of *sound.*

Mach's Principle. The *inertia* of a body due to its interaction with the rest of the *Universe*. An isolated body has no inertia.

Macro. Large.

Macrobian. Long-lived.

Macrobiotics. Art or science of prolonging life.

Macrocyclic. Containing a ring structure consisting of more than tweleve *atoms* in the *molecule.*

Macrometer. Optical instrument provided with focussing telescope and two mirrors with which ranges of distant objects can be estimated.

Macromolecular. Consisting of or pertaining to *macromolecules.*

Macromolecule. A very large *molecule*, a *polymer.*

Macronucleus. Larger of the two nucleus in ciliate protozoa.

Macronutrient. Essential element that are needed by plant in relatively large amount for its normal growth *e.g.,* carbon, hydrogen, iron, magnesium, nitrogen etc.

Magellanic Clouds. Two small patches of *light* which appear, from the southern hemisphere, to be detached from the main bright band of *stars* constituting the *Milky Way.*

Magenta. Fuchsine $C_{20}H_{22}N_3OCl$. A red dye can be prepared from *aniline* and *toluidine*

Magic Number. 2, 8, 20, 28, 50, 82 and 126—atomic nuclei containing these numbers of protons have exceptional stability.

Magma. Molten material, consisting of *silicates* with occluded *gases* forms the mantle or crust of the *Earth.* Magma given out by *volcanoes* is called lava.

Magnadur. *Ferrite*, used for making permanent *magnets.*

Magnalium. A light *alloy*, of aluminium containing 5% to 30% magnesium and 1% to 2% copper.

Magnesia. *Basic magnesium carbonate*; A *solution* of magnesium hydrogencarbonate.

Maenesite. *Magnesium carbonate*, $MgCO_3$, occurs in white masses: used in the manufacture of *refractories* and *fertilizers.*

Magnesium (Mg). Silvery-white metal, exists in *nature* as carnallite, dolomite, epsom salt, magnesite etc., boiling point 1107°C, extracted by electrolysis of fused magnesium chlorides; used in flush bulbs, for synthesis of many organic compounds, extraction of silicon etc.

Magnesium Bromide ($MgBr_2$). White crystalline substance, obtained by heating magnesium in bromine vapour; found in sea-water.

Magnesium Carbonate ($MgCO_3$). White powder occurs in nature as magnesite and dolomite; used in medicines as mild antacid.

Magnesium Chloride ($MgCl_2$). Anhydrous salt obtained on heating hexahydrate in current of dry hydrochloric acid gas: used as thread lubricant in spinning and weaving industry.

Magnesium Hydroxide [Mg $(OH)_2$]. White powder, slightly soluble in water, used in sugar industry extract to sugar from molasses.

Magnesium Iodide (MgI_2). White hydroscopic salt obtained by dissolving magnesium oxide in hydroiodic acid; it easily decomposes liberating iodine.

Magnesium Nitride (Mg_3N_2). Greenish-yellow amorphous powder obtained when magnesium is burnt in atmosphere of nitrogen; reacts with water forming megnesium hydroxide and ammonia $Mg_3N_2 + 6H_2O \rightarrow 3Mg(OH)_2 + 2NH_3$.

Magnesium Oxide (MgO). White powder soluble in water, obtained when magnesium is burnt in air $2Mg + O_3 \rightarrow 2MgO$; used for refractory lining of furnaces, making crucibles and also in medicines.

Magnesium Perchlorate [Mg $(ClO_4)_2$]. Obtained by dissolving magnesium hydroxide in perchloric acid; used as drying agent.

Magnesium Sulphate ($MgSO_4$). Found in nature as kieserite ($MgSO_4 . H_2O$), well defined transparent crystals, lose water of crystallisation on exposure to air; used as purgative, mordant, in manufacture of paints, soaps etc.

Magnesothermic Reduction. *Reduction of oxides* to the corresponding metals at high temperatures using metallic *magnesium.*

Magnet Permanent. A *ferromegnetic substance* having a permanent *magnetic field* and *magnetic moment.*

Magnetic Amplifier. A device for the amplification of small *direct currents* and of low *frequency alternating currents.*

Magnetic Bottle. Configuration of *magnetic fields* used in the *containment* of a *plasma in* controlled *thermonuclear reaction* experiments.

Magnetic Bubble. A *computer* memory element.

Magnetic Circuit. A closed path that follow the *line of force* of a *magnetic field.*

Magnetic Constant. Permeability of free space. μ_o. It has the value $4\pi—10^{-7}$ henry per metre.

Magnetic Declination. The *angle* between the *planes* of the geographic and *magnetic merian.*

Magnetic Dip. The *angle* between the direction of the Earth's magnetic field and the horizontal.

Magnetic Disk. A metal disc coated with magnetic *iron oxide*, used as a *computer* memory.

Magnetic Domains. Regions, within a *ferromagnetic* material, in which the atomic *magnetic moments* are aligned when the material is magnetized.

Magnetic Elements. The three quantities, *magnetic declination, magnetic dip*, and *horizontal component*, which define completely the Earth's magnetic field at any point.

Magnetic Equator. Aclinc line. A line of zero *magnetic dip.* It lies fairly near the geographical *equator.*

Magnetic Field. A *field* of *force* that exists at any point. A magnetic field may exist at a point as a result of the presence of either a permanent *magnet* or a circuit carrying an *electric current.*

Magnetic Field of Electric Current. A wire or coil carrying an *electric current* is surrounded by a *magnetic field.*

Magnetic Field Strength. Magnetic intensity. *H.* The strength of a *magnetic field* measured in *amperes* per *metre* (*SI units*) or *oresteds* (*c.g.s. units*).

Magnetic Flux. Φ. The strength of a *magnetic field* through an area. *SI unit* of magnetic flux is the *weber*.

Magnetic Flux Density. Magnetic induction. *B*. The *magnetic flux* passing through unit area of a *magnetic field* in a direction at right angles to the *magnetic force*. *SI unit* of magnetic flux density is the *tesla* (*weber* per square *metre*.)

Magnetic Force. The *force* exerted by a *magnetic field* on a *magnetic pole*.

Magnetic Induction. The induction of *magnetism* in a body by some external *magnetic field*.

Magnetic Line of Force. The path along with a free *magnetic pole* would be accelerated.

Magnetic Mirrors. The regions of high field strength at the end of an externally generated *magnetic field* used in the *containment* of a *plasma* in a controlled *thermonuclear reaction* experiments.

Magnetic Moment. The torque experienced by a magnetic *dipole* in a field of unit *magnetic fields strength* perpendicular to it. Measured in weber metres. Also called the magnetic dipole moment.

Magnetic Monopole. A hypothetical unit of magnetic 'charge' analogous to *electric charge*.

Magnetic Permeability μ**.** The ratio of the *magnetic flux density* in a medium to the external *magnetic field strength* that induces it.

Magnetic Pole. A *magnet* appears to have its *magnetism* concentrated at two points known as the poles.

Magnetic Storm. A sudden disturbance in the Earth's magnetic field it affects *compasses* and *radio* transmission.

Magnetic Susceptibility X_m. The ratio of the *magnetization* (*M*) produced in a substance to the *magnetic field strength* (*H*) to which it is subjected.

Magnetic Tape. *Plastic* tape coated with a *ferromagnetic* powder, used in tape recorders.

Magnetism. The branch of physics related with *magnets* and *magnetic fields*.

Magnetism Terrestrial Geomagnetism. The Earth's magnetism.

Magnetite. Magnetic iron ore. Black *iron oxide*, Fe_3O_4.

Magnetization M. The magnetic moment per unit volume of a magnetized body.

Magneto. A small *alternating—current dynamo* having a secondary winding to produce a high voltage that enables a *spark* to jump between the *electrodes* of a sparking plug during the ignition of a petrol engine.

Magnetohydrodynamics (MHD). The study of behaviour of moving electrically conducting fluids in magnetic fields.

Magnetometer. Instrument used for comparing strengths of *magnetic fields* and *magnetic moments.*

Magnetomotive Force (MMF). Magnetic potential. It is the circular integral of the *magnetic field strength* around a closed path.

Magneton. A *unit* for measuring the *magnetic moments* of atomic particles.

Magnetosphere. The *space* surrounding the *Earth,* or any other celestial body, in which there is a *magnetic field* associated with that body.

Magnetostriction. Refers to change in the dimensions of *ferromagnetic substances* on magnetization. It arises due to changes in the sizes of the *magnetic domains.*

Magnetron. *A thermionic valve* capable of producing high power oscillations in the *microwave* region.

Magnification. The ratio of the length of the image to the length of the object.

Magnifying Glass. *Microscope.*

Magnifying Power of a Compound Microscope. The ratio of the *angle* substended at the eye by the final image to the angle subtended by the object placed at the least distinct vision.

Magnifying Power of a Lens. The ratio of the *angle* subtended at the eye by the *virtual image* to the angle subtended by the object when placed at the least distance of distinct vision.

Magnitude of Stars. It is a measure of the relative apparent *brightness* of *stars.*

Magno. Alloy containing 95.5% nickel and 4.5% manganese: used in manufacture of incandescent lamps and radio tubes.

Magnox. A magnesium *alloy* used for sheathing uranium *fuel elements* in *nuclear reactor.*

Major Axis. The *axis* of an *ellipse* passing through both *foci.*

Major Gene. Gene that governs readily identifiable expression of phenotype.

Majority Carriers. The *carrier* that constitutes more than half the total number of carriers in a semiconductor.

Maksutov Telescope. An astronomical telescope. It consists of a *concave* spherical *mirror* whose *aberration* is reduced by a *mineral lens.*

Malachite. *Basic* copper carbonate, $CuCO_3Cu(OH)_2$. A bright green *mineral* used as a gemstone and as a copper *ore.*

Malate. A *salt* of *malic acid.*

Malathion ($C_{10}H_{19}O_6PS_2$). Organo-phosphate, brownish liquid, boilling point 156°C, soluble in most organic solvents, unstable in alkaline medium; one of the safest insecticide, contact and stomach poison.

Maleate. Salt of maleic acid.

Maleic Acid. Cis-butanedioic acid.

Maleic Hydrazine. A herbicide, inhibits growth or sprouting.

Mallgnant. Growth and poliferation of some neoplasm, it is likely to prove fatal if not checked by medical treatment.

Malic Acid. 2-hydroxybutanedioic acid. $HOOCCH(OH)CH_2COOH$. A white crystalline *organic acid,* m.p. 98-99°C. Occurs in unripe apples and other fruits.

Malleability. It is the ability to be hammered out into thin sheets.

Malonic Acid. *Propanedioic acid.*

Malonyl. The *bivalent radical* —$OCCH_2CO$— drived from *propanedioic* (malonic) *acid.*

Malonylurea. *Barbituric—acid.*

Malt. Germinate gram is heated and dried to get malt.

Maltase. An *enzyme* present in *yeast* and other *organism,* it hydrolyzes *maltose* into *glucose.*

Maltose. Malt sugar, maltobiose. $C_{12}H_{22}O_{11}$. A hard crystalline *soluble disaccharide,* formed in *malt* by the *enzyme amylase* on *starch.*

Malt Sugar. *Maltose.*

Mandelic Acid ($C_5H_5CHOHCOOH$). A white crystalline *optically active* substance, used as an *antiseptic.*

Manganese (Mn). Greyish-white, hard, brittle metal found in nature as oxides (MnO_2), boiling point 2040°C; used mainly in making alloys and steel.

Manganese Bronze. Alloy containing 55 to 60% copper, 38 to 42% zinc and upto 4% manganese.

Manganese Dioxide (MnO_2). Black powder, insoluble in water, decomposes into Mn_3O_4 and oxygen; used as catalyst, in manufacture of dry cells and in glass industry.

Manganese Steel. Steel containing upto 13% manganese. It is very hard.

Manganin. Alloy containing 80 to 85% copper, 12 to 15% manganese and 2 to 4% nickel; used for resistance coils.

Manganous Chloride ($MnCl_2 . 4H_2O$). Pink crystals obtained by dissolving manganese (oxide, hydroxide) in hydrochloric acid; can also be prepared by heating manganese in chlorine; used as catalyst.

Manganous Hydroixide [$Mn(OH)_2$]. White precipitate obtained by adding alkali solution to solution of manganous salt. On exposing to air it immediately becomes brown.

Manganous Oxide (MnO). Greyish-green powder, obtained by reducing higher oxide with hydrogen or carbon monoxide at 1200°C.

Manganous Nitrate [$Mn(NO_3)_2$]. Colourless deliquescent crystals, obtained by dissolving manganous carbonate in dilute nitric acid.

Manganous Sulphate ($MnSO_4$). Obtained on heating manganese dioxide (MnO_2) with sulphuric acid.

Mannitol [$HOCH_2(CHOH)_4CH_2OH$]. White solid, optically active polydric alcohol; used in manufacture of synthetic resins and plasticisers.

Mannitol Hexanitrate [$C_6H_8(ONO_2)_6$]. Colourless, insoluble in water used in medicine and as an explosive.

Manometer. Instrument used for determining pressure of a gas.

Margarine. Butter substitute, prepared from hydrogenated oils.

Mariner's Compass. Compass with needle always pointing north-south; used by navigators.

Markownikoff (Markownikov) Rule. During addition of a hydrogen *halide* to an asymmetric *alkene* the *halogen atom* becomes attached to the carbon atom with the lesser number of hydrogen atoms.

Mars. A *planet*, with two small *satelites*, having its *orbit* between those of the *Earth* and *Jupiter*.

Marsh Gas. *Methane*, CH_4.

Marsh's Test. A sensitive test for arsenic. It depends upon the formation of *arsine*.

Martensite. Hard and brittle steel obtained when material is cooled from its hardening temperature at much faster rate than its normal cooling rate.

Mascon. A local concentration of high *mass*, below the surface of the *Moon*.

Maser. Microwave Amplification by Stimulated Emission of Radiation. Masers can also be made to operate at optical frequencies, when they are referred to as optical masers or *lasers*.

Mass Action Law. The velocity of a chemical change is proportional to the *active masses (molecular concentrations)* of the reacting substances.

Mass Decrement. The difference between the *rest mass* of a radioactive *nuclide* and the rest masses of its *decay products*.

Mass Defect. The difference between the *rest mass* of a *nucleus* and the sum of the rest masses of its constitutent *nucleons*.

Mass-energy Equation. The equation $E=mc^2$, where c is the speed of *light* in ms^{-1} and E is the energy, in *joules*, released when a mass m, in *kilograms*, is completely converted into energy. In *c.g.s., units*, if m is in *grams* and c is in $cm\ s^{-1}$, E will be in *ergs*.

Massicot. A yellow powder *lead oxide*, PbO.

Mass Number. The number of *nucleons* in the *nucleus* of an *atom*.

Mass Spectrograph. An apparatus used for the determination of the exact *masses* of individual *atoms*.

Mass Spectrometer. An apparatus for obtaining the *mass spectrum* of a *beam* of *ions* by means of suitably disposed *magnetic* and *electric fields*.

Mass Spectrum. A *spectrum* obtained with a *mass spectrometer* or *spectrograph* in which a *beam* of *ions* is arranged in order of increasing *charge* to *mass* ratio.

Masurium. Old name of element technetium (Tc).

Mating. Meeting or coming together of individuals of opposite sex for sexual reproduction.

Mating Strain. Group of organisms of a species which do not interbreed with each other but breed with members of other group of same species which are morphologically identical but genetically and physiologically different.

Matrix. Intercellular material in which animal cells are embeded.

Matte. A mixture of the *sulphides* of iron and copper obtained in the *smelting* of copper.

Matter. A form of *energy* that has the attributes of *mass* and extension in *space* and time.

Mauve. Mauveine, aniline violet. A reddish-violet *dye*; it was the first organic dye prepared artificially.

Maxillae (Sing. Maxilla) Pair of large bones of upper jaw of vertebrates; in mammals, each bears molar and premolar teeth. Also found in mouthparts of insects and crustaceans.

Maxilullae. First pair of maxillae in crustaceans.

Maximum. (Math.) A *function* $y=f(x)$ has a maximum value at $x = a$ if $f(a)$ is greater than the values of the function immediately preceding and immediately following $x = a$.

Maxwell. The *c.g.s. units of magnetic flux.* The flux through 1 square *centimetre normal* to a *magnetic field* of strength 1 *gauss.* 1 maxwell = 10^{-8} *weber.*

Maxwell-Boltzmann Distribution. A statistical *equation* giving the distribution of *veocities* of the *molecules* in a *gas.*

Mechanics. Branch of science that deals with behaviour of bodies when these are subjected to forces.

Mean. Average.

Mean Free Path. The average, or *mean,* distance travelled by a particle, *atom,* or *molecule* between collisions.

Mean Free Time. The average, or *mean,* time between two collisions of a particle, *atom,* or *molecule.*

Mean life. The reciprocal of the *disintegration constant.*

Mechanical Advantage. The ratio of the actual load raised to the *force* required to maintain the machine at constant speed.

Mechanical Equivalent of Heat. If *H* units of *heat* are completely converted into *W* units of *work* then $W=JH$, where J is a constant called the mechanical equivalent of *heat.*

Mechanics. The branch of science dealing with the behaviour of *matter* under the action of *force.*

Median. 1. A line joining a vertex of a *triangle* to the mid-point of the opposite side. **2.** The middle number in a sequance of numbers.

Median Lethal Dose (M) LD.$_{50}$ The *dose* of *ionizing radiation* that would kill 50% of a large batch of *organisms* within a specified time.

Medium Frequencies. M.F. *Radio frequencies* in the range 300-3000 *kilohertz.*

Medulla. The central part of certain bodily organs, such as the adrenal glands, in which the medulla functions separately from the cortex, which surrounds it.

Meerschaum. Natural *hydrated* magnesium silicate, $Mg_2\ Si_3O_8.2H_2O$, white *solid,* used for tobacco pipes.

Mega. Prefix denoting one million times, in *metric units*; symbol M.

Meghertz. MHz. 1 million *hertz* equal to 10^6 *cycles* per *second.*

Megaton Bomb. A *nuclear weapon* with an explosive power equivalent to one million tons of *T.N.T.* approximately 4×10^{15} *joules.*

Megohm. One *million ohms.*

Meiosis. The process by which the *nucleus* of a *diploid* reproductive *cell* divides itself to produce four *haploid* cells —the *gametes.*

Meissner Effect. The reduction in the *magnetic flux* within a *superconductor,* when it is cooled below its critical temperature in a magnetic field.

Melamine. $C_3H_6N_6$. Triaminotriazine. A white crystalline substance, m.p. 354°C., forms a *thermosetting* resin with *formaldehyde.*

Melanin ($C_{17}H_{98}O_{33}N_{14}S$). A dark brown pigment produced in the skin *cells.* Skin and hair colours in many animals, including man, are due to melanin. The function of the melanin is to absorb the harmful radiations of the sun.

Melting. Process by which solid is turned into liquid.

Melting Point. Temperature at which solid and liquid coexist.

Membrane. Tissue surrounding cells, organs, structure etc., mostly semipermeable allowing only selected substances to pass through.

Mendelevium (Md). At No. 101 transuranic element Md^{285} is most stable of its isotopes and its *t* 1/2 = 60 days.

Mensuration. Refers to measurement of lengths, areas and volumes.

Menthol ($C_{10}H_{19}OH$). Colourless crystals. They have strong peppermint odour and cooling taste; used as antiseptic in ointments. in medicines and perfume industry.

Meobal ($C_{10}H_{13}NO_2$). White crystalline solid; used as insecticide.

Mephosfolan ($C_8H_{16}O_3S_2PN$). Yellow poisonous liquid, boiling point 120°C, soluble in organic solvents; used as acaricide and insecticide.

Mercaptan (RSH). Thio-alcohol containing sulphur in place of oxygen.

Mercaptide. *Thiolates.*

Mercuric. A *compound* of mercury in its + 2 *oxidation state e.g.* $HgCl_2$ mercuric chloride.

Mercurous. A *compound* of mercury in its + 1 *oxidation state, e.g.* Hg_2Cl_2 mercurous chloride.

Mercury (astr.). A *planet,* orbit nearest the *sun,* having no atmosphere and a day temperature of about 400°C.

Mercury. Quicksilver, hydrargyrum. Hg. Element. R. a.m. 200.59. At. N. 80 A *liquid*, silvery-white *metal*, r.d. 13.6, m.p.-39°C., b.p. 357°C. It occurs as *cinnabar*, HgS. Extracted by roasting the *ore* in a current of air. It is used in *thermometers, barometers, manometers,* etc. *alloys* (called *amalgams*) are used in denstistry. *Compounds* are used in medicines.

Mercury Cell. A *primary cell.* The *E.M.F.* is about 1.3 *volts.*

Mercury Chlorides. 1. Mercurous chloride, calomel, Hg_2Cl_2. A white *insoluble* power, used in medicine and as a *fungicide.* **2.** Mercury (II) chloride, mercuric chloride, corrosive sublimate $HgCl_2$. A poisonous white *soluble salt,* used as an antiseptic and in preparation of other mercury compounds.

Mercury Cyanate. Mercuric fulminate. $Hg(ONC)_2$. A white crystalline substance, explodes on being struck, used in detonators to initiate explosions.

Mercury (II) Oxide. HgO. A *soluble* poisonous powder, occurs as yellow or red *crystals*; used as a pigment and as an *antiseptic.*

Mercury (II) Sulphide. HgS. An *insoluble* substance, occurs naturally as *cinnabar* used as a pigment, known as *vermilion.*

Mercury-vapour Lamp. A lamp emitting a strong bluish *light ultraviolet radiations*; used in artificial sun-ray treatment and in street lighting.

Meridian, Celestial. The *great circle* of the *celestial sphere* passing through the *zenith* and the celestial poles. It meets the horizon at points known as North and South points.

Meridian, Terrestrial. Meridian of longitude. An imaginary *great circle* drawn round the *Earth* passing through both poles.

Mescaline ($C_{11}H_{17}NO_3$). White soluble powder, obtained from mescal (peyote cactus); used as hallocigen.

Mesitylene [$C_6H_3(CH_3)_3$]. Colourless hydrocarbon, boiling point 164 to 165°C; occurs in coal-tar, used in organic synthesis.

Mesityl Oxide [$(CH_3)_2C:CHCOCH_3$]. Colourless oily liquid, boiling point 130 C; used as solvent.

Meso. A prefix indicating that the compound is optically in active.

Mesoderm. Refers to that Germ layer from which muscles, connective tissues and blood system usually develop.

Mesomerism. *Resonance.*

Mesons. A group of unstable *particles* belonging to *hadrons* class. Positive, negative, and neutral mesons exist. Mesons are found in *cosmic rays.*

Mesophases. *Phases* that are intermediate between crystalline and liquid phases.

Mesosphere. Earth's *atmosphere* between the *ionosphere* and the *exosphere,* exending from about 400 kilometres to 1000 kilometres above the Earth's surface.

Mesyl. Methylsulphonyl. The *univalent radical* CH_3SO_2—

Meta. 1. Denoting 1, 3 position in *benzene ring*. Abbreviated to *m*-as a prefix in naming a compound; *e.g. m*-dichlorobenzene is 1, 3-dichloro-benzene. **2.** A prefix indicating an *inorganic acid* (or a corresponding *salt*) of a lower degree of hydration; *e.g. metaphosphoric acid,* HPO_3, as compared with *orthophosphoric acid,* H_3PO_4.

Metabolism. The chemical processes associated with living *organisms.* Metabolic reactions are under the control of *enzymes,* which are of immense importance in the chemistry of life.

Metabolite. Refers to any substance that takes part in the process of metabolism.

Metacrate ($C_9H_{11}O_2N$). White crystalline substance, dissolves in acetone and alcohol but insoluble in C_6H_6, $C_6H_6 . CH_3$ and H_2O. It is used as an insecticide.

Metagenesis. It is the alteration of sexual and asexual reprocdction in life cycle of certain animals.

Metal. A substance having a 'metallic' lustre and being malleable, ductile, of high *relative density,* and a good *conductor* of *heat* and *electricity.*

Metaldehyde. Meta. A white, *volatile,* flammable poisonous solid. It is a *polymer* of *ethanal* (acetaldehyde) CH_3CHO used as *fuel* in small heaters and as a slug killer.

Metallic Crystals. The type of *crystal* formed by most *metals.* In which the outer *electrons* of the *metallic atoms* are shared by the crystal as a whole.

Metallic Soap. An *insoluble salt* formed by a *metal* and a *fatty acid,* used for waterproofing textiles and as a drier for *paints.*

Metallocene. *Ferrocene.*

Metallography. It relates to study of the crystalline structure of *metals* and *alloys.*

Metalloid. *Element* having some properties of *metals* and others of non-metals. An element giving rise to an *amphoteric oxide.*

Metallurgy. The science of the extraction of metals from their *ores.*

Metamerism. A type of *isomerism* exhibited by *organic compounds* of the same chemical type; caused by the attachment of different *radicals* to the same central atom or group.

Metamict Minerals. *Minerals* whose crystalline structure has been disrupted by *alpha particles* produced by *radioactive nuclei* within the minerals.

Metamorphic Rock. Rock, formed from existing rock, by heat, pressure, or chemical shown by fluids.

Metaphosphoric Acid. HPO.

Metastable State (chem.). The state in which stable *phase* is not formed unless a small amount of the normally stable phase is introduced. Thus supercooled water will remain as *liquid* water below 0°C until a small *crystal* of *ice* is introduced.

Metastable State (phys.). An excited state (see *excitation*) of an *atom* or *nucleus* having an appreciable life-time.

Meteor. A *solid* body from outer *space.*

Meteorology. The science of the weather; observing atmospheric *pressure, temperature, wind* strength, *humidity,* etc., to draw conclusions regarding the forthcoming weather.

Meteor Showers. Exceptionally heavy falls of *meteors.*

Meter. Suffix denoting measurer; *e.g., voltmeter.*

Methacrylate. A *salt* of *methacrylic acid.*

Methacrylic Acid. 2-methylpropenoic acid. $CH_2{:}C(CH_3)COOH$. A corrosive *liquid,* m.p. 15°C., b.p. 163°C. Methyl methacrylate, is an important *plastic.*

Methanal. Formaldehyde. HCHO. The simplest *aldehyde.*

Methane (CH_4). Non-poisonous, colourless, odourless gas, occurs in gases in oil- wells; used for making hydrogen, methylchloride, methylene chlorine, methanol and formaldehyde.

Methanoat Formate. A *salt* of *methanoic acid* (formic acid).

Methanoic Acid. Formic acid. HCOOH. A colourless, corrosive fuming liquid with a pungent smell, m.p. 84°C., b.p. 100.5°C. It occurs in various plants and in ants; simplest of the *carboxylic acids*, used in dyeing, tanning, and electroplating.

Methanol. Methyl alcohol, wood spirit. CH_3OH. A colourless, poisonous liquid, b.p. 65.6°C., obtained by the *destructive distillation* of wood. Used as a *solvent,* and in the chemical industry.

Methionine. An *amino acid* found in *casein*, wool, and other *proteins*, used in the treatment of liver diseases.

Methoxy. The *univalent radical*, CH_3O—.

Methoxy Benzene. Anisole methyl phenyl ether. $CH_3OC_6H_5$. A colourless liquid with an aromatic odour, b.p. 155.4°C., used in perfumes and as a vermicide.

Methyl. The *univalent* organic *radical* CH_3 —.

Methyl Alcohol. *Methanol.*

Methylmine. CH_3NH_2. A *gas* with an odour of *ammonia*, b.p.—6.4°C.

Methylated Spirit. A *liquid fuel* consisting, by *volume*, of 90% *ethanol*, 9.5% *methanol* 0.5% *pyridine*, together with small amounts of *petroleum* and some methyl violet dye.

Methylated Spirit, Industrial. A variety of *methylated spirit* free from *pyridine*, it consists of *ethanol* with 5% *methanol*.

Methylbenzene. Toluene, toluol. $C_6H_5CH_3$. A hydrocarbon of the *benzene* series. A colourless flammable *liquid* with a characteristic odour, b.p. 110°C. Occurs in *coal-tar* can be made from methylcyclohexane from *petroleum*, used as a solvent and in the manufacture of *T.N.T.*, *saccharin* and *drugs*.

Methyl Chloride. *Chloromethane.*

Methyl Cyanide. *Acetonitrile.*

Methylcyclohexanol. $CH_3C_6H_{10}OH$. A colourless *liquid* b.p. in the range 167 - 174°C; obtained from *methylphenol*, used as a *solvent* for *rubber* and *cellulose*.

Methylene. *Carbene.*

Methylene Blue. $C_{16}H_{18}N_{13}SCl$. A *soluble* , intense blue *dye*. Used as a dyestuff, in medicine, as a stain in *biology*.

Methyl Isocyanate (MIC). Reactive, very toxic and volatile gas, used in manufacture of carbamate pesticides (carbaryl etc.); its leakage at Bhopal (December 2, 1984), caused death of over 2000 people.

Methyl Methacrylate. *Methacrylic acid; polymethyl methacrylate.*

Methylol. Hydroxymethyl. The univalent radical $OH.CH_2$—.

Mathyl Orange. $C_{14}H_{14}N_3NaO_3S$. An orange *indicator*, used in acid-base titrations. It is red relow a *pH* of 3.1 and yellow above 4.4.

Methyl Parathion ($C_8H_{10}NO_5PS$). White solid, less stable than ethyl parathion ($C_{10}H_{14}NO_5PS$), highly poisonous to insects; used as insecticide.

Methyl Red. Acid dye prepared by coupling diazotised o-amino benzoic acid with dimethylaniline; used as indicator in acidalkali titration.

Methyl Salicylate. Oil of wintergreen, methyl 2-hydroxybenzenecar boxylate. $OH.C_6H_4COOCH_3$. A colourless oil, b.p. 223.3°C., used in flavours, perfumes and medicine.

Methyl Sulphate [$CH_3)_2$ SO_4]. Poisonous, colourless syrupy liquid, immiscible with water; used in laboratory for introduction of methyl radical in alcohols, amines and phenols.

Metol. 4-methylaminophenol. **$CH_3NH.C_6H_4OH$.** A white *compound*, m.p. 87°C. Used as a developer in *photography.*

Metre m. The *SI Unit* of length, one metre is equal to 39.3701 inches.

Metre Bridge. *Wheatstone bridge.*

Metre-candle. *Lux.*

Metric System (units). A system of *weights* and measures originally based upon the *metre.*

Metric Ton. Tonne. 1000 *kilograms*; 2204.61 Ib, 0.9842 ton.

Metrology. The study of weight and measures.

MeV. Million *electron-volts.*

Mho. Reciprocal ohm. The unit of electrical *conductance*, known as the *siemens.*

Mica. A group of *minerals*, the most important of which are muscovite, $H_2KAl_3(SiO_4)_3$, and phlogopite, $H_2KMg_3Al(SiO_4)_3$. An excellent insulator and resistant to hgih *temperatures*, used as a *dielectric* in *capacitors*, as a support for *electrodes* in *thermisonic valves*, and for heating elements in irons, etc., it is used for inspection windows of furnaces being transparent.

Micelle. A cluster or group of associated *molecules*, found in a *colloidal solution.*

Michelson-Morley Experiment. An attempt to measure the *velocity* of the Earth through the *ether*.

Micro. Very small. One millionth (10^{-6}).

Microanalysis. Chemical analysis of minute quantities.

Microbalance. Balance capable of weighing very small weights of the order of 10^{-3} to 10^{-6} g.

Microbial insecticide. Insecticide of pathogenic microbes or their toxic products.

Microbicide. Toxin capable of killing microbes.

Microchemistry. Chemistry relating to microscopic or sub-microscopic quantities or objects.

Microcomputer. A small personal computer using a *microprocessor* as its control unit.

Microcosmic Salt. *Ammonium sodium hydrogen orthophosphate.*

Microcrystalline. Having crystalline structure which can be seen only under microscope.

Microelectronics. The design, Manufacture, and use of *electronic* units utilizing extremely small *solid-state* components, generally based on *integrated circuits* and silicon *chips.*

Microfared μF. One-millionth of a *farad.*

Micrometeorite. A *meteor* with a diameter of less than 1 mm.

Micrometer. Instrument for the accurate measurement of small distances or *angles.*

Microminiaturization. The techniques used in *microelectronics.*

Micron (μ). Thousandth of millimetre.

Micronutrient. Nutritive substance or trace element required in minute quantity by plant.

Microphone. Instrument that picks up and changes sound-waves into electrical current, for amplifying intesifying or transmitting.

Microphotograph. Photograph taken using a microscope; photograph reduced to microscopic size.

Microscope. Instrument that produces magnified images of small objects by use of lens (es).

— **simple,** fitted with one simple convex lens only, magnifying power limited.

— **compound,** having two converging lenses-eye-piece and objective, for greater magnification.

— **electron,** using parallel beam of electrons, to achieve magnification upto 250000 times.

— **ultra-violet,** using ultra-violet rays to obtain magnification beyond 1500 times; quartz lenses are used and image is recorded photographically.

Microtome. Apparatus used for cutting very thin sections of material, for microscopical examination.

Microwave. In wireless communication, having very short wavelength.

Microwave Spectroscopy. The measurement of the absorption or emission of *electromagnetic radiation* in the *waveband* 0.1 mm to 10 cm by atomic or molecular systems.

Mil. One thousandth of an inch.

Milk. The fluid secreted by female mammals to provide food for their offspring.

Milk of Lime. A *suspension* of *lime* in *water.*

Milk of Magnesia. Magnesium hydroxide [$Mg(OH)_2$], gives milky white suspension in water; used as laxative.

Milky Way. The luminous band of stars encircling the heavens. These stars are members of the *Galaxy.*

Millepore. Hydrozoan coral with many pores.

Milli. Prefix denoting one thousandth. Symbol m.

Milliammeter. A sensitive *ammeter* can measure *milliamperes.*

Milliampere. MA/ 1/1000 *ampere.*

Millibar. A unit of atmospheric pressure, used in *meteorology.* 1000 *dynes* per square centimetre or 100 *pascals* approximately equal to 1/32 inch of mercury.

Milligram. Mg. 1/1000 *gram*; 0.0154 *grain.*

Millilitre. Ml. 1/1000 *litre.*

Millimetre. Mm. 1/1000 *metre*; 0.0393701 inch.

Mineral. A naturally occuring inorganic substance.

Mineral Oil. Oil of mineral origin. It is a mixture of hydrocarbons *e.g.,* kerosene oil.

Minicomputer. A *computer* which could be packed in a single desk-top cabinet.

Minim. Fluid measure; 1/60 of a fluid drachm; 0.0591 cm^3.

Mink. Small stoat-like mammal (several species), about 50 to 70 cm long; its fur is quite valuable.

Mirror. Looking glass; smooth surface that reflects images (real or virtual). Concave mirror forms real image of object only when it is placed at distance greater than its focal length.

Misch-Metal. Alloy of lanthanum, cerium and didymium; used for electrodes, in gas lighters and tracer bullets.

Mispickel (FeAsS). Mineral composed of iron, arsenic and sulphur; sulfarsenide of iron.

Mist. Droplets of *water,* formed by the *condensation* of *water-vapour* on dust particles.

Mitochondria. Minute rod-shaped bodies, about 1-3 μ m in diameter occur in the *cytoplasm* of cells. It contain many of the *enzymes* of the cell, specially those required by the *citric acid cycle.*

Mixture. Heterogenous association of two or more substances mixed together in any proportions the constituents can be separated by suitable physical means.

Moa. Extinct, herbivorous bird.

Module. One-package components which operate as complete functional unit in system or subsystem.

Modulus. A constant *factor* for the conversion of *units* from one system to another.

Moebius Strip. A ribbon-shaped strip of paper one end of which has been twisted through 180 degrees before it is attached to the other end.

Moho. Mohorovicic discontinuity between the *Earth's* crust and its underlying mantle.

Mohs Scale of Hardness. A scale in which each *mineral* listed is softer than all those below it. 1. *Talc.* 2. *Gypsum.* 3. *Calcite.* 4. *Fluorite.*

Molality (m). Number of moles of solution present in one kg of solvent.

Molar (s). Grinding teeth; permanent back teeth of mammals.

Molarity (M). Number of moles of solute present in one litre of solution.

Molar Solution. A *solution* with a *concentration* of 1 *mole* per dm^3.

Molar Volume. Molecular volume V_m. The volume occupied by 1 *mole* of a substance. At 760 mmHg and $0^\circ C$, the molar volume of a *perfect gas* is 22.415 dm^3 per mole.

Mole. *SI unit* of amount of substance. 1 mole of a compound has a mass equal to its *relative molecular mass* in grams. 1 mole of electrons has a mass of 5.486×10^{-4}g. It replaces such former units as the *gram-atom, gram-molecule, gram-ion,* and *gram equivalent,* Symbol *mol.*

Molecular Biology. Relates to the study of the structure of the *molecule,* that are of importance in *biology e.g.,* proteins etc.

Molecular Compound. Chemical *compounds* formed by the chemical combination two or more complete *molecules.*

Molecular Distillation. *Evaporation* of *molecules* from a surface, at 10^{-2} mm kg *pressures* and their *condensation* under such conditions that their *mean free path* is of the same order as the distance between the heated and cooled surfaces. Used for *isotope separation.*

Molecular Formula. *Formula* of a chemical *compound.* It shows the kind and the actual number of *atoms* present in the *molecule.*

Molecular Sieves. Highly porous *aluminosilicate adsorbents,* containing pores of uniform size. They are used as drying agents and in the purification of fluids.

Molecular Spectrum. The *spectrum* emitted by *molecules.* Caused by transitions between various states of molecular rotation, vibration. etc.

Molecular Weight. *Relative molecular mass.*

Molecule. Smallest unit of element or chemical compound capable of free existence.

Molluscide. Toxicant used in killing slugs and snails.

Molybdenum (Mo). Hard, silvery-white metal, boiling point 4800°C, Found in nature as molybdenite ($MoSO_4$) and wulfenite ($PbMoO_4$); used in manufacture of alloy steels and in making filaments of electric lamps.

Molybdenum Trioxide (MoO_3).—Yellow crystalline substance; used to manufacture molybdenum compounds.

Molybdic Acid (H_2MoO_4). Yellow crystalline substance, loses a molecule of water at 70°C and forms molybdenum trioxide.

Monacid (Monacid). Having one replaceable hydrogen atom.

Monad. An element having a valency of one.

Monazite. Phosphate Mineral of cerium, thorium and other rare-earths with some occulated helium.

Mond Process. Used in the extraction of nicked by the action of *carbon monoxide,* CO, on the impure *metal.*

Monel Metal. *Alloy* of copper (25% - 35%), nicked (60% - 70%) and iron, manganese, silicon, and carbon, used as an *acid* resisting material in chemical industry.

Monobasic Acid. An *acid* having one *atom* of *acidic* hydrogen in a *molecule; e.g., nitric acid,* HNO_3.

Monoblepsis. Condition wherein vision is highly distinct when only one eye is used.

Monochromatic. Having one colour or wave-length only.

Monocle. Single eye-glass.

Monoclinic. Relates to crystal shapes, having three unequal axes with one oblique intersection.

Monocompound. Compound having only one atom or group.

Monocrotophos ($C_7H_{14}O_5NP$). Colourless crystals having mild ester smell, miscible with acetone, alcohol and water; used as systemic insecticide.

Monocular. One-eyed.

Monocyclic. Containing one whole or ring.

Monohydrate. Having one molecule of water.

Monohydric. Having one hydroxyl group.

Monolayer. A layer one molecule thick.

Monoplane. Aeroplane or glider having one set of wings.

Monoplanetic. Fungi that produces only one type of planospores.

Monopode. One-footed man.

Monosaccharide. Simple sugar which cannot be hydrolysed to lower carbohydrates— fructose, glucose etc.

Monotropic. Capable of existing in only one stable form.

Monovalent. Univalent.

Monoxide. Oxide containing one oxygen atom in molecule.

Moped. Light motorized bicycle.

Mordant. Inorganic compound that is used for dyeing cloth etc.

Morphine ($C_{17}H_{19}O_3N$). Chief alkaloid found in opium obtained from poppy; morphine hydrochloride is used in medicine as a soporific and as pain releiver.

Morpholine [O (CH_2CH_2)NH]. Colourless, hygroscopic substance boiling point 128°C; used as solvent for resins and waxes.

Morphology. Refers to study of forms and features of different parts.

Mortar. A building material consisting mainly of *lime* and *sand* that hardens on exposure through chemical action between the ingredients and atmospheric *carbon dioxide.*

Mosaic Gold. Crystalline stannic sulphide (SnS_2) made of shining golden-yellow scale.

Mother-Liquor. Solution left after formation of crystals.

Mucilage. Plant product obtained from roots, seeds or other plant parts using extraction with cold or hot water.

Mucin. Glycoprotein which lubricates food, found in saliva, gastric secretion and large intestinal secretion.

Mucoproteins. Proteins containing carbohydrate group.

Multicellular. Containing more than one cell.

Multienzyme System. Sequence of related enzymes that participate in a given metabolic pathway.

Multimeter. Instrument used for measuring various electrical quantities such as currents, resistance, voltages etc.

Multiple Resistance. Resistance of plant to several stress such as diseases, drought, insects, mites, nematodes etc.

Multiple Star. A system of *stars* made up of three or more components held together by *gravitation.*

Multiplet. **1.** A line in a *spectrum* formed by two or more lines, they result from small differences of *energy level* in the atoms or molecules. **2.** A group of related elementary particles differing only in electric *charge.*

Multiplication Constant (factor). For *nuclear* reactor is the ratio of the average number of *neutrons* produced by *nuclear* fission per unit time, to the total number of neutrons absorbed or leaking out in the same time.

Multiplicity. The number of *energy levels* into which an *atom* or *nucleus* splits.

Multivalent. Refers to group of more than two homologues formed during meiosis in polyploids or polysomics.

Mumetal. Alloy with high magnetic permeability containing upto 78% nickle in addition to iron, copper and manganese.

Muntz Metal. Alloy, Cu and Zn in ratio of 3 : 2; used when strong brass is required.

Muon. Elementary particle having a mass 207 times that of an electron.

Muriate. Old term for chloride.

Muriatic Acid. Old term for hydrochloric acid.

Mustard Gas. Dichlorodiethyl sulphide. $(CH_2CH_2Cl)_2S$. An oily *liquid* used as a war gas, destroyed by *oxidizing agents, e.g., bleaching powder.*

Mutagan. Substance that produces *mutations.*

Mutarotation. A change in optical rotation.

Mutation. Sudden inheritable divergence from ancestral type.

Myrcene ($C_{10}H_{16}$). Colourless oil, found in verbena and bay oil.

Myrioscope. Type of kaleidoscope.

Myristic Acid [$CH_3(CH_2)_{12}$ COOH]. One of the fatty acid components found in human fat.

Myrmecochry. Disemination of seeds or other reproductive material by ants.

Myrrh. Bitler-tasting, sweet-smelling gum or resin obtained from shrubs; used in making incense and perfumes.

N

NAA (Naphthalene Acetic Acid). Synthetic auxin, useful in stimulating and promoting flowering.

NAD (Nicotinamide Adenine Dinucleotide). Coenzyme that serves as hydrogen accepter; it picks up hydrogen, forming dihydronicotinamide anenine dinucleotide (NADH).

Nadir. The lowest point. The point opposite the zenith on the celestial sphere.

NADP (Nicotinamide Adenine Dinucleotide Phosphate). Coenzyme with similar action as NAD.

Naiad. Aquatic gill breathing nymyh, common in many insect species.

Nails. Flattened horny plate at the end of finger or toe, present in many primates, including man.

Name. A means of referring to or identifying some element in a computer program in a computer network or in some other system.

Nano. One thousand millionth. 10^{-9}.

Nanometre. Unit of length equal to one thousandth of micrometre; millimicron. 1 Nm = 10^{-9} m.

Napalam Bomb. Incendiary bomb that contains highly inflammable petroleum jelly, causes fires that are difficult to extinguish.

Naphtha. It is a mixture of hydrocarbons in various proportions obtained from paraffin oil, coal-tar etc.

Naphthalene ($C_{10}H_8$). Colourless lustrous substance, insoluble in water but soluble in hot ethanol, cold ether, benzene etc., used as repellent to avoid attack by insect pests.

Naphthenes. Saturated hydrocarbons having at least one closed ring of carbon atoms.

Narceine ($C_{23}H_{27}NO_8 3H_2O$). White crystalline alkaloid, occurs in opium; used as muscle relaxant.

Narcotic. Any drug that relieves pain and induces sleep; large doses may cause stupor, coma or even death.

Nascent Hydrogen. New born hydrogen, obtained by electrolysis of water or from acids by action of metals. It possess enormous intrinsic energy; used for preparing hydrides of antimony, arsenic and phosphorus.

Natrium. *Sodium.*

Natron. Natural sodium sesquicarbonate, $Na_2CO_3.NaHCO_3.2H_2O$.

Natural (Chem.). Occurring in nature.

Natural Abundance. The *abundance* of each different *isotope* in an *element* as it is normally found in nature.

Natural Frequency. Frequency at which a system vibrates in absence of any external force.

Natrual Gas. Mixture of gaseous hydrocarbons, mostly methane, coming out of earth.

Natural Glass. Amorphous, vitreous inorganic material that has solidified from magma too quickly to crystallise.

Natural Radioactivity. Radioactivity exhibited by naturally occurring radionuclides.

Natural Science (s). Sciences that deal with objectively measurable phenomena pertaining to relationship and transformation of energy and matter—botany, chemistry, physics, zoology etc.

Nautical Mile. Geographical mile; one minute (1/60 degree) of longitude measured along equator—1854 metres (6082.66 feet); used by sailors in navigation.

Near Infrared or Ultraviolet. The shortest *infrared* or longest *ultraviolet wavelengths*; *i.e.*, those wavelengths of these two type of *radiation* that are 'nearest' in magnitude to those of visible light.

Neel-Point. Temperature at which magnetic susceptibility becomes normal.

NEFA (Non Essential Fatty Acids). Fatty acids present in plasma and are bound to lipoproteinic complexes.

Negative Lens. Cancave lens.

Negative Pole. South-seeking pole of magnet.

Nekton. Refers to assemblage of actively swimming animals in lake, sea or ocean waters.

Neodymium (Nd). Metal of rare-earth group.

Neon (Ne). Monoatomic colourless, odourless, inert gas. Present in air in small quantity (0.002%), boiling point $-246^{\circ}C$., obtained by fractional distillation of liquid air; used in neon signs and lights.

Neon Tube. A *discharge* tube containing neon at a low pressure.

Neoprene. $(CH_2 = CH—C(Cl) = CH_2)_n$—A synthetic rubber, has high tensile strength, heat and ozone resistance better than ordinary rubber.

Neotype. A specimen selected to serve as type specimen when all original material is lost.

Neper. A unit for expressing the *ratio* of two values (*e.g., currents, voltages,* etc.) equal to the natural *logarithm* of the ratio of the quantities. 1 neper =8.686 *decibels.*

Nephelometer. Instrument for measuring turbidity of *liquids,* or *scattering of light* by particles in *suspensions.*

Nephelometry. Method of quantitative analysis. It involves spectrophotometric estimation of scattering of light by colloidal suspension.

Nephoscope. A grid like instrument used to determine the speed of celestial objects by observing time of transit.

Neptune (astr.). A *planet* with two *satellites. Its orbit* lies between those of *Uranus* and *Pluto.* The surface temperature is about 200°C and the dense atmosphere consists mainly of *methane* and *hydrogen.*

Neptunium. Np. *Transuranic element.* At. No 93. Most stable *isotope,* neptunium-237, has a *half-life* of 2.2×10^6 years. A *metal* of silvery appearance, r.d. 20.45, m.p. 640°C., produced as a *by- product* in the manufacture of plutonium.

Nernst Effect. If a *temperature* gradient is maintained across an electrical *conductor* (or *semiconductor*) that is placed in a transverse *magnetic field,* a *potential difference* will be produced across the conductor.

Nernst Heat Theorem. The entropy change for chemical reactions involving crystalline *solids* is zero at the *absolute zero* of temperature.

Nerol ($C_{10}H_{17}OH$). Colourless unsaturated alcohol, isomeric with geraniol, boiling point 224°C, obtained from neroli oil; used in perfumes.

Neroli Oil. Essential oil extracted from flowers of orange trees.

Nerve Gas. A gas used in wars. It attacks the Nervous system. Generally these gases are derivatives of phosphoric acid.

Nesseler's Solution. Solution of potassium mercuric iodide ($KHgI_3$) in potassium hydroxide solution; used as test for ammonia, it forms brown coloured precipitate with ammonia.

Neutral (Chem.). Neither acidic nor alkaline having a pH of 7.

Neutralization (chem.). The addition of *acid* to *alkali*, or *vice-versa*, till *solution* is *neutral*.

Neutral Temperature. The temperature of the hot junction of a *thermocouple* at which the *E.M.F.*, is a maximum and the rate of change of E.M.F., with *temperature* is zero.

Neutrino. A stable *elementary particle* with no *electric charge* or *rest mass*, but with *spin* $^1/_2$.

Neutron. It has no *electric charge* and a *mass* slightly greater than that of the *proton* (1.67492×10^{-27} *kilogram*).

Neutron Flux. A measure of the number of *neutrons* passing through unit area in unit time, *i.e.*, the number of neutrons per unit volume multiplied by their mean speed.

Neutron Number *N*. The number of *neutrons* in an atomic *nucleus*; it is equal to the *mass number* minus the *atomic number*.

Neutron Star. A hypothetical state of a *star* at the end of its evolutionary process when it has consumed all its *nuclear fuel* and no longer has a source of internal *energy*. Although, it is thought that *pulsars* may be this type of star.

Neutron Temperature. The *energies* possessed by *neutrons* in thermal equilibrium expressed in terms of a *temperature*, if they behave as a *monatomic gas*. The neutron temperature on the *Kelvin scale*, *T*, is given by; $E=3kT/2$, where E is the neutron energy and k is *boltzmann's constant*.

New Candle. *Candela.*

Newt. Any of very small, tailed amphibian.

Newton (N). *SI* unit of force, equal to force required to accelerate one kilogram by one metre second^{-2}.

Newtonian Telescope. Newtonian reflector. A form of astronomical *telescope*.

Newton's Law of Cooling. The rate at which a body loses heat is proportional to the *temperature* difference between the body and its surroundings.

Newton's Law of Gravitation. Every particle in the *Universe* attracts every other particle with a *force* directly proportional to the product of the *masses* of the particles and inversely proportional to the square of the distance between them. Thus, the force of attraction between two masses m_1 and m_2, separated by a distance s, is given by:

$$F = Gm_1m_2/r^2,$$

where G is the *gravitational constant*.

Newton's Law of Motion. The following laws are known as Newtones laws.

— **First,** a body continues in its state of rest or of uniform motion in straight line unless compelled by external force to change that state.

— **Second,** rate of change of momentum is proportional to impressed force and takes place in direction of force.

— **Third,** to every action there is an equal and opposite rection.

Newton's Rings. Coloured rings that may be observed round the point of contact of a *convex lens* and a *plane* reflecting surface. Caused by the *interference* effects that occur between *light* waves reflected at the upper and lower surfaces of the air film separating the lens and the flat surface.

Niacin. Vitamin (nicotine acid); its deficiency causes pellagra.

Nichrome. Alloy of nickel, chromium and iron having high melting point and high resistivity; used for making electric resistors and heating elements.

Nickel (Ni). Silvery-white magnetic metal, malleable and ductile, can be easily forged, rolled and welded, resistant to air and moisture, boiling point 2840°C; used in electroplating and to make stainless steel.

Nickel Acetate [$(CH_3COO)_2$ $Ni.4H_2O$]. Green soluble substance; used in nickel-plating.

Nickel Carbonate ($NiCO_3$. $6H_2O$). Green solid obtained by adding sodium bicarbonate solution to nickel salt solution, insoluble in water but soluble in acids; used in ceramic and in electroplating.

Nickel Carbonyl [$Ni(CO)_4$]. Colourless liquid obtained by passing carbon monoxide over finely divided nickel at 60°C; used as catalyst.

Nickel Chloride ($NiCl_2.6H_2O$). Green hexahydrate got by dissolving metal in aqua regia or by action of dilute hydrochloric acid on nickelous oxide or carbonate.

Nickel Dioxide (NiO_2). Black monohydrate obtained by passing chlorine through suspension of nickel hydroxide in water.

Nickelite Oxide (Ni_2O_3). Black powder obtained by heating nickelous oxide (NiO) in air, insoluble in water, dissolves in concentrated hydrochloric acid.

Nickel Nitrate [$Ni(NO_3)_2 6H_2O$]. Green hydrate obtained by dissolving metal oxide or carbonate in dilute nitric acid and crystallising the product.

Nickelous Hydroxide [$Ni(OH)_2$]. Green precipitate obtained by heating nickelous salt solution with caustic soda.

Nickelous Oxide (NiO). Greenish powder obtained by heating nickelous hydroxide or carbonate; insoluble in water but soluble in acids; reduced to nickel on heating in hydrogen $NiO+H_2 \rightarrow Ni+H_2O$.

Nickel-Plating. Depositing thin layer of nickel on another metal by electrolysis.

Nickel Silver (German silver). Alloy containing about 60% copper, 20% nickel and 20% zinc.

Nickel Steel. Steel containing upto 9% nickel.

Nickel Sulphate ($NiSO_4 . 7H_2O$). Green crystals formed by dissolving carbonate in dilute sulphuric acid; used in ceramics, paints, varnishes etc., and as mordant in dyeing and in nickel-plating.

Nicol Prism. Optical device constructed from crystal of calcite; used to obtain plane polarised light.

Nicotinamide. Part of NAD molecule that participated in various biological reactions.

Nicotine ($C_{10}H_{14}N_2$). Colourless oily liquid that is readily soluble in water; pure nicotine has unpleasant odour unlike tobacco.

Nictitating Membrane. Third eye-lid; transparent, than membrane possessed by many vertebrates that opens and closes laterally across cornea.

Night Temperature. Temperature prevailing around organisms (plant or animal) during period of darkness; this in many plants greatly affects the growth— low night temperature lowers the rate of respiration.

Niobirum (Nb). Rare grey metal, discovered in mineral tantalite, earlier named columbium; small quantities in stainless steel preserve steel's corrosion resistance at high temperature.

Nit. Unit of luminance equal to one candela per square metre.

Niton. Old name for Radon.

Nitrate. A salt of nitric acid.

Nitrating Acid. Mixture of nitric and sulphuric acids; used for nitration.

Nitration. Reaction in which nitro group is introduced into organic compound.

Nitric Acid (HNO_3). Colourless fuming liquid having sharp chocking smell, readily miscible in water, boiling point 80°C; used in drugs, dyes, lacquers, plastics and explosives.

Nitrification. Process of Oxidation of nitrogen-containing ions in soil by bacteria.

Nitrile. Organic compound containing—CN group (cyanide); generally a liquid with pleasant smell.

Nitrile Rubber. Copolymer of butadience with acronitril; It is used in sealing gaskets, hoses, rings etc.

Nitrobenzene ($C_6H_5NO_2$). Pale yellow oily liquid smelling like bitter almonds, boiling point 211°C, used as solvent, oxidising agent and in manufacturing aniline, azodes, floor polishes etc.

Nitrocellulose. Pulpy cotton-like amorphous solid contains 10 to 14% nitrogen; used in fast drying lacquers, rocket propellant, high explosives etc.

Nitro-chalk. Mixture of calcium carbonate ($CaCO_3$) and ammonium nitrate (NH_4NO_3); used as fertiliser.

Nitro Compound. Organic compound having nitro group attached to an aromatic ring.

Nitrogen (N). Colourless, odourless and tasteless gas found in atmosphere (4000 billion tonnes) as free nitrogen, essential for growth of animal and plant life; used in manufacture of ammonia, nitric acid, nitrides etc.

Nitrogenase System. System of enzymes that is capable of reducing atmospheric nitrogen into ammonia in presence of ATP.

Nitrogen Balance. Difference between nitrogen (protein) intake and nitrogen excretion for an individual. It is positive in growing children and negative in patients having wasting diseases.

Nitrogen Cycle. Circulation of nitrogen and its compounds between organism and environment.
The cycle is given below.

Nitrogen Dioxide (NO_2). Formed by reduction of nitric acid and by action of heat on some nitrates; used as an oxidising and nitrating agent.

Nitrogen Fixation. Reaction in which atmospheric nitrogen is changed into useful nitrogen compounds.

Nitrogenise. To combine or supply with nitrogen.

Nitrolime. A mixture of $CaCN_2$ and C; used as a fertilizer.

Nitrogen Monoxide (Nitric Oxide, NO). Colourless gas formed by action of dilute nitric acid on copper turnings ; used as reducing agent.

Nitroglycerine. Colourless poisonous oily liquid formed by adding glycerol in thin stream to cold mixture of concentrated nitric acid and sulphuric acid.

Nitrometre. Apparatus used for estimating nitrogen or some of its compounds.

Nitromethane (CH_3NO_2). Colourless oily liquid, boiling point 101°C., used as solvent.

Nitrosation. Conversion of ammonium salts into nitrites.

Nitrourea ($NH_2CONHNO_2$). Colourless crystalline solid, obtained by addition of urea nitrate to concentrated sulphuric acid; used in preparation of semicarbazide.

Nitrous Acid (HNO_2). Weak acid of nitre; used to manufacture dyes.

Nitrous (Dinitrogen) Oxide (N_2O). Colourless, non-inflammable gas obtained on heating ammonium nitrate upto 270° ; used as anaesthetic and in aerosols.

Nobelium At. No 102. NO^{255} is most stable, $t_{1/2}$ = 3 min.

Noble Gas (es). Group zero elements, monoatomic, inert and rare gases. They are helium (He), neon (Ne), argon (Ar), krypton (Kr), xenon (Xe) and randon (Rn); occur in atmosphere in very small quantities, randon is produced by radioactive decay ; these atoms show no tendency to combine with each other to form molecules.

Noble Metal (s). Metals which do not corrode or tarnish in air or water and are not easily attacked by acids *e.g.*, gold, platinum and silver.

Noble Prize (s). Initiated by Swedish scientist, Alfred Noble, to be awarded annually in chemistry, literature, peace, physics and medicine, economic science was added in 1969. Indian Noble laureates are Rabindra Nath Tagore in literature (1913) and C.V. Raman in physics (1930).

Nocturnal. Active during night.

Nocturnal Enuresis. Sleep disorder that results in bed wetting.

Nocuous. Injurious, poisonous.

Nodal Point. Optical centre of eye.

Node. Point in stationary wave pattern at which vibration is minimum. Place at which leaf is attached to stem, often swollen. Knob or knot.

Noise. Unmusical sound ; cacophony.

Nomogram. Nomograph. An alignment chart by which the value of a *variable* can be found, without calculation, from the values of one or two other known variables.

Nonanoic Acid. Pelargonic acid, $CH_3(CH_2)_7$ COOH. A colourless oily *liquid,* b.p. 253-5°C., used in the manufacture of *lacquers* and *plastics.*

Non-aqueous Solution. Any solution in which solvent is not water but may be organic or inorganic, protonic or non-protonic.

Non-electrolytes. Substance that does not conduct electricity in aqueous solution ordinary conditions.

Non-ferrous Metal. Any metal other than iron or steel.

Non-homologous Pairing. Association of non-homologous segments of one or two chromosomes at pachytene.

Non-metallic Elements. Any electronegative element whose oxide can form acids; these may be solid (carbon), liquid (bromine) or gaseous (nitrogen).

Non-Newtonian Fluid. A fluid that does not obey Newton's law of viscosity, *i.e.*, the viscosity is not independent of the rate of shear or the velocity gradient. In *colloids* and other fluids consisting of more than one *phase* the viscosity usually diminishes as the velocity gradient increases.

Non-Ohmic. Circuit component, current passing through it does not obey Ohm's law.

Non-persistant. Pesticides that breaks quickly and become inactive.

Non-polar Compound. Compound having no permanent dipole moment *e.g.*, benzene, carbon tetrachloride etc.

Non-preference. Behaviour of animals to avoid a plant or host species.

Non-secretor. Person with A and B antigens on red blood corpuscles but not having in any body secretion.

Nor-adrenaline. Lacking methyl group present in adrenaline.

Normal. Having a straight chain of carbon atoms.

Normality. Number of gram equivalent weights of solute in one litre of solution.

Normal Solution. Solution having one gram equivalent of dissolved substance in a litre of solution.

Normal Spectrum. Spectrum having spectral lines separated by angles that are proportional to difference in their wave-length.

Notation. A system of symbols for representation of numbers, quantities, or other entities.

Nova. A *star* ejecting a small part of its material in the form of a *gas* cloud. During the process the star becomes more luminous than it was before the outburst.

NTP (Normal Temperature and Pressure). Normal temperature is 273 K and pressure 76 cm of mercury.

Nuclear Energy. Energy obtained from nuclear fission or nuclear fusion.

Nuclear Fission. Nuclear reaction in which a heavy atomic nuclei breaks into two nearly equal fragments with release of very large amount of energy.

Nuclear Force. Strong attractive force that exists between nucleons when they are extremely close together (closer than 10^{-15} m).

Nuclear Fuel. Fissionable or fertile isotope having reasonably long half-life that undergoes nuclear fission or nuclear fusion in nuclear reactor.

Nuclear Fusion. Nuclear reaction in which two light nuclei fuse together to form heavy nucleus releasing a very large amount of energy.

Nuclear Physics. Branch of physics dealing with structure and properties of nuclei, nuclear reactions and nucleon collisions .

Nuclear Reaction. Reaction, involves a change in nucleus of an atom.

Nucleon. Refers to protons and neutrons which are constituents of all nuclei.

Nucleon Number. Number of nucleons in atomic nucleus; mass number.

Nucleoplasm. Substance of nucleous except chromatin material. It contains complexes and enzymes necessary for replication of DNA and synthesis of RNA molecules.

Nucleosidase. Enzyme that act on nucleosides breaking union between purine or pyrimidine combined with pentose sugar.

Nucleoside. Any one of base sugar complexes which are formed when purine or pyrimidine base combines with carbon-1 of pentose sugar.

Nucleotidase. Enzyme that acts on nucleotides releasing phosphoric acid resulting in formation of nucleosides.

Nucleotide. Molecule consisting of phosphate, pentose sugar and purine or pyrimidine.

Nucleus (pl. Nuclei). Round body of protoplasm in cell containing chromatin and producing enzymes that stimulate activities of cytoplasm.

Nuclide. Nucleus of an atom containing a given number of nucleons —O^{16} and O^{17} are different nuclides.

Nyctalopia. Visual abnormality; caused by deficiency of vitamin A and leading to night blindness.

Nylarder Reagent. A Solution of potassium sodium tartrate, potassium or sodium hydroxide and bismuth subnitrate in water; It is used for testing sugar in urine.

O

Obease. Abnormally fat.

Obesity. Pathological condition in human being. It is characterised by excessive deposition of fat in subdermis, resulting in increase in body weight and dimensions, may lead to diabetes etc.

Object. The source of *rays* affected by an optical instrument to form an image.

Objective. Lens or combination of lenses used in microscopes.

Object Program. The program that is output by a complier or assembler and is ready to be loaded into a computer for execution.

Obligate Parasite. Parasite that is not capable of free existence *e.g.*, liverfluke, tapeworm etc.

Obligate Saprophyte. Saprophyte that cannot survive in absence of dead organic matter.

Oblique. Shape of leaf having two unequal halves.

Occultation. The cutting off of the *light* or *radio* emission from one celestial body when another is interposed between it and the observer. *e.g.*, a *star* may become invisible to an *optical* or *radio telescope* when it is hidden behind the *Moon.*

Oceanography. Relating to the study of the oceans, sea floor, sea waters, and their tides and currents.

Occlot. A type of American cat. It resembles a small leopard.

Ochre. A natural *hydrated* form of Fe^2O^3, containing various impurities used as a red or yellow *pigment.*

Octa-, Octo-. Prefix denoting eight.

Octacalcium Phosphate OCP. $Ca_2H(PO_4)_3$. $2.5H_2O$. A crystalline substance, Important in the chemistry of bones, teeth.

Octadecanoic Acid. Stearic acid. $C_{17}H_{35}COOH$. A white *fatty acid* m.p. 69°C., occurs as *glycerides* in many fats, used in the manufacture of soaps and cosmetics.

Octagon. An eight-sided *polygon.*

Octahedron. A *polyhedron* with eight faces.

Octane. $CH_3(CH_2)_6CH_3$. Colourless *liquid,* b.p. 126°C., r.d. 0.704, occurs in *petroleum.* The isomer 2,2,4-trimethylpentane (iso-octane) also occurs in *petroleum.*

Octane Number of a Fuel. The percentage by *volume* of iso-*octane* in a mixture of iso-octane and *normal heptane,* C_7H_{16} that is equal to the *fuel* in knock characteristics under specified test conditions.

Octanoic Acid. Caprylic acid. $CH_3(CH_2)_6COOH$. A colourless oily liquid, b.p. 237°C., used in the manufacture of *dyes* and perfumes.

Octanol. Octyl alcohol. $C_8H_{17}OH$. Refers to a group of *isomeric alcohols* of which the most important is *I*-octanol, a colourless *liquid,* m.p. 16.7°C., b.p. 194—9°C., used as a *solvent.*

Octant. The portion of a *circle* cut off by an arc and two radii at 45; one-eighth of the area of a circle.

Octavalent. Having a *valence* of eight.

Octave. The interval between two musical notes.

Octaves, Law of. An incomplete statement of the *periodic law* made by J.A.R. Newlands in 1864.

Octet. It is the group of eight electrons constituting outer electron shell of inert gases (except helium) or atom.

Octo. Eight.

Octogenarian. Person having age over eighty years (below ninety).

Octyl. Univalent radical, C_8H_{17}—.

Ocular. Relating to eye.

Oculist. One who is specialist in diseases of eye.

Oculomotor. Third cranial nerve of vertebrates it supplies eye muscles and is responsible for rotation of eye-ball.

Odd-even Nucleus. Nucleus containing odd number of protons and even number of neutrons.

Odd-odd Nucleus. Nucleus that has odd number of protons and odd number of neutrons.

Odometer. Instrument used for recording distance covered by vehicles.

Oersted. The C.G.S. unit of *magnetic field strength* or *magnetic intensity* defined as the strength of a magnetic field that causes a unit *magnetic pole* to experience a *force* of 1 *dyne* in a vacuum. It is equivalent to $1/4\pi \times 10^3$ *amperes* per *metre.*

Ohm (Ω). Unit of electrical resistance. Resistance having potential difference of one volt when one ampere current is passed through it.

Ohmmeter. A multimeter used to measure R.

Ohm's Law. Ratio of potential difference between ends of a conductor and current flowing through it.

Oil Cake. A mass of oilseeds from which the *oil* has been expelled in a press or extracted by a *solvent* used as cattle food.

Oil of Vitriol. *Sulphuric acids.*

Oil of Wintergreen. *Methyl salicylate.*

Oil Sand. *Sandstone* or loose sand impregnated with a viscous *hydrocarbon* mineral oil.

Oil Shale. A fine-grained *sedimentary rock* containing kerogen, a form of organic matter that decomposes on heating to yield mineral oil.

Oil, Synthetic. Natural *mineral oils* composed of various *hydrocarbons.* It is possible to make similar products artificially from *coal,* etc.

Olber's Paradox. If the Universe contains an infinite number of uniformly distributed stars the night sky should be uniformly bright. In fact, it is not: this is explained by the *expansion of the Universe* and the recession of the galaxies, which causes a *redshift* making the most distant galaxies no longer visible. Named after Heinrich Wilhelm Olbers (1758-1840).

Oleate. A *salt* of *oleic acid.*

Olefiant Gas. Old name for *ethene.*

Oleic Acid. Cis-octadec-9-enoic acid. $C_{17}H_{33}COOH$. An *unsaturated liquid organic acid,* m.p. 15°C., occurs in the form of *glycerides* in many *fats* and *oils*. A high proportion of *triolein.*

Oleoyl. The *univalent unsaturated radical* $C_{17}H_{33}CO$—(form *oleic acid*).

Oleum ($H_2S_2O_7$). Colourless fuming liquid obtained by dissolving sulphur tri-oxide in concentrated sulphuric acid; on diluting with water, it becomes sulphuric acid again.

Oligomenorrhea. Abnormality in menstruation when there is scanty menstruation.

Oligomer. Polymer made up of comparatively few monomer units in molecule.

Oleyl Alcohol. $(CH_3(CH_2)_7CH{:}CH(CH_2)_7CH_2OH$. An *unsaturated liquid alcohol,* b.p. 205°C., used in organic synthesis.

Olivine. $(Mg, Fe)_2SiO_4$. A mineral *silicate* of magnesium and iron. Used as a gem.

Omega-minus ∩$^-$. A negatively charged *elementary particle*, known as a *hyperon* and having a mass 3276 times that of the *electron*.

Omegatron. Instrument in which *ions* are caused to move in spiral path under the influence of an *electric field* at right angles to a constant *magnetic field*. The instrument may be used for the absolute determination of atomic *masses* and for isotopic and chemical analysis.

Omnivore. Eating every thing. Animal which feeds or lives on both animal and vegetative food.

Onchogenic Virus. Virus which can induce cell proliferation and tumour formation.

Oncology. Relating to the study of cancer, its causes and cure.

Oncotic Pressure. Osmotic pressure due to plasm colloids.

Ontogeny. History of development of an individual from egg cell onwards.

Opacity. It is the *reciprocal* of the *transmittance.*

Opal. Hydrated *amorphous silica*, SiO_2, used as gem.

Opaque. Not permitting a *wave motion* (*e.g., light, sound , X-rays*) to pass. A medium that does not permit *X-rays* or *gamma-rays* to pass is said to be *radio-opaque.*

Open-chain Compounds. *Organic compounds* that are not derived from ring compounds.

Open Clusters. Clusters of *stars* having a common motion through *space.*

Open-hearth Process. Siemens-Martin process. A process for *steel* manufacture. *Pig-iron* and steel scrap or *iron ore* in calculate amounts are heated together by *producer gas* on a hearth in a furnace.

Operator. A symbol representing a mathematical operation to be carried out on a particular operand.

Operon. A group of *genes* control the *synthesis* of the individual *enzymes* that act together as one enzyme system.

Ophthamology. It is the study of the eye and its diseases and the correction of impaired vision.

Ophthalmoscope. Instrument for examining the eye. A powerful light and *lens system.*

Opium. The dried, milky juice from unripe fruits of the opium poppy, *papaver somniferum*. It contains several *alkaloids*, including *morphine, narceine* and *codeine.*

Opsin. Protein forming photosensitive compound present in human eye and other mammals.

Optic. Relating to sight.

Optical Activity. Property of some substances to rotateplane of vibration of plarised light.

Optical Axis. Imaginary straight line that passes through centre of curvature of simple lens.

Optical Centre. A point, situated at the geometrical centre of a thin *lens,* through which an incident *ray* passes without suffering any deviation.

Optical Fibres. A *glass* fibre that functions as a *waveguide* for light. Used in medical instruments to examine internal organs also used in short-range *telecommunications.*

Optical Glass. *Glass* used in optical instruments.

Optical Isomerism. Enantimorphism. The occurrence of a compound in two different forms, one a mirror image of the other.

Optically Flat. A surface is said to be optically flat if the irregularities do not exceed the *wavelength* of light.

Optical Telescope. An astronomical *telescope* used to observe celestial bodies.

Optical Temperature. The *radiation temperature* of a celestial body as calculated from its *light radiation.*

Optic Axis. The direction in a doubly refracting *crystal* in which *light* is propagated without *double refraction.*

Optic Chiasma. X-shaped structure formed by crossing of optic nerves in mid brain.

Optic Lope (s). Thickened part of mid brain connected with sight.

Optic Nerve (s). Nerves from eye to brain.

Optics. Relates to stuuy of nature and properties of sight.

Optic Stalk. Narrow passage formed by constriction of proximal portion of optic vesicle; serves to attach eye-ball.

Optometer. Instrument used for sight-testing.

Orbit. Closed path of moving objects; commonly used for path of moon around earth, path of planets around sun.

Orbital. One that moves in an orbit.

Order (Bio.). Major catagory in taxonomie hierarchy. It comprises of families.

Order (math.). The number of times a *function* has been differentiated to give a particular *derivative.*

Order to Magnitude. A magnitude expressed to the nearest power of 10.

Ordinary Ray. When a *ray* of *light* is incident upon a *crystal* which exhibits *double refraction* so that the direction of the ray makes an angle with the *optic axis* of the crystal, the ray splits into two rays. One of these obeys the ordinary laws of *refraction* and is called the ordinary ray.

Ordinate. In *analytical geometry*, the ordinate of a point is the perpendicular distance of the point from the *x*-axis.

Ore. A naturally occurring mineral material from which a desired product can be extracted; *e.g.*, *bauxite* is an ore of aluminium.

Organic Acid. Organic compound, it is liable to give up proton to a base phenol.

Organic Base. Molecule or ion possessing a lone pair of electrons which can be used for coordination with a proton.

Organic Chemistry. Branch of chemistry that deals with hydrocarbons and their derivatives.

Organochlorine Compound (s). Compounds having carbon, hydrogen and chlorine as molecular constituents – some common insecticides like BHC ($C_6H_6Cl_6$), DDT ($C_{14}H_9Cl_5$) etc.

Organometallic Compounds (s). Colourless mobile liquids having low boiling points; the metal is directly linked to carbon.

Organ Pipe. Air column enclosed in a pipe having one or both ends open.

Ormulu. Alloy of copper, tin and zinc in various proportions, with at least 50% copper.

Orgenesis. The building of a mountain.

Orpiment. Natural *arsenic trisulphide*, As_2S_3. A yellow mineral.

Orsat Apparatus. A portable apparatus used for determining the amount of *carbon dioxide*, oxygen, and *carbon monoxide* in exhaust *gases*.

Ortho. Denoting adjascent position in a hexagonal ring of atoms, particularly the *benzene ring*. Abbreviated as *o*-as a prefix in naming a compound, *e.g.*, *o*-dichlorobenzene is 1, 2- dichlorobenzene.

Orthochromatic Film. A photographic *film* that is sensitive to green to green in addition to blue and violet *light*.

Orthoclase Feldspar. Natural potassium aluminium silicate, $K_2O.Al_2O_3.6SiO_2$. A constituent of *granite*.

Orthogonal. 1. Rectangular, or involving right angles. **2.** Having a set of mutually *perpendicular axes*.

Orthohydrogen. Hydrogen *molecules* in which the *spins* of the two *nuclei* are parallel.

Oscillation. Regularly repeated to-and-fro motion about a point.

Oscillograph. Apparatus used for producing a curve, representing a wave-form.

Osmiridium. Natural alloy of osmium and iridium having smaller amount of platinum, rhodium and ruthenium; hard and resistant to corrosion; used for tipping pen-nibs.

Osmium Os. Element. R.a.m. 190.2. At. No. 76. A hard white crystalline *metal.* The heaviest substance known; r. d. 22.57, m.p. 3045°C, b.p. 5027°C. Its occurs together with platinum *(see osmiridium)* and is used in *alloy* with platinum and iridium.

Osmometer. Apparatus used for measuring osmotic pressure.

Osmosis. Diffusion of fluids or solutions through porous membrane.

Osmotic Pressure. Pressure due to differences of concentration in solutions present on opposite sides of semi-permeable membrane.

Osprey. Large bird of prey, belongs to hawk family, that feeds on fish etc.

Ostwald's Dilution Law. A law relating the dissociation constant, K and the degree of dissociation α, of a weak *electrolyte* of *concentration c moles* per *litre*. It states that for a binary electrolyte.

$$K=ca^2/(1-a).$$

Otoscope. Instrument used to examine the ear.

Otto Engine. A four-stroke *internal-combustion engine* invented by Nicolaus Otto (1832-91) in 1876.

Ouabain $C_{29}H_{44}O_{12}$. A white crystalline *glycoside*, m.p. 200°C.,obtained from wood; used as a heart stimulant.

Ounce. 1. 437.5 grains or 28.3 grams. **2.** 8 fluid drachms or 28.41 cm^3. **3.** 480 grains or 31.1 grams.

Ovshinsky Device. Ovonic device. A device consisting of a special *glass*, made up of selenium and tellurium, whose *resistance* drops rapidly when a suitable *voltage* is applied across it. These are used as special purpose switches in *electronic circuits.*

Oxalic Acid ($C_2H_2O_4.2H_2O$). Poisonous, colourless, crystalline solid soluble in water but insoluble in ether; used in manufacture of ink.

Oxidant. Any substance capable of giving oxygen to bring about combustion *e.g.*, hydrogen peroxide or liquid oxygen in rocket fuels.

Oxidase. Any of a group of enzyme which promote oxidation in animal and plant cells.

Oxidation. Process of addition of negative ions or removal of positive ions from molecule : dehydrogenation.

Oxidative Decarboxylation. Oxidation by removal of carbon dioxide and hydrogen.

Oxidative Phosphorylation. Process of addition of inorganic phosphate to ADP by one energy-rich bond to form ATP.

Oxide. Binary compound of oxygen.

Oxidising Acid. Any acid acting as oxidising agent *e.g.*, nitric acid, sulphuric acid.

Oxidoreductase. Any enzyme that catalyses oxidation as well as reductions.

Oxime. Organic substance containing C. NOH group, obtained by reaction of aldehyde or ketone with hydroxal-amine (NH_2OH).

Oximetry. Process of measurement of oxygen saturation of circulating blood.

Oxirane (Epoxyethane). Colourless gas, obtained on oxidation of ethane in presence of catalyst; used for making ethanediol, ethanolamines etc.

Oxonium Ion. Old name of hydronium ion (H_3O^+).

Oxydiaacetic Acid [$O(CH_2COOH_2)$]. White soluble dibasic organic acid; used in manufacture of plastics and plasticisers.

Oxygen (O). Colourless and odourless gas, occurs in atmosphere (21%) as free oxygen ; water contains about 89% by weight of combined oxygen, essential for respiration used in breathing apparatus.

Oxygenation. Refers to temporary combination of oxygen to haemoglobin at respiratory surface.

Oxygen Carrier. Molecule that can react directly with molecular oxygen to form peroxide or superoxide derivative and can subsequently use oxygen in oxidation reaction *e.g.*, haemoglobin.

Oxygen Debt. Accumulation of lactic acid in muscles during violent exercise, due to *oxygen* shortage.

Oxyhaemoglobin. Unstable chemical compound in blood formed on combination of oxygen with haemoglobin ; when it goes to tissue, oxygen is released due to dissociation of compound.

Oxyntic Cell (s). Acid (HCl) secreting cells of stomach ; parietal cells.

Oxytocin. Hormone of neurophypophysis its main action is on milk ejection in lactating animals.

Ozone (O_3). Bluish gas, chemically very active, obtained by passing oxygen through silent electric discharge ; used as bleaching agent and germicide also used for purifying air and water.

Ozonide. A compound formed by action of ozone on unsaturated olefinic compounds.

Ozonisation. Process of combining with ozone.

Ozoniser. Apparatus used to convert oxygen into ozone by subjecting oxygen to silent electric discharge.

Ozonolysis. Addition to ozone to unsaturated hydrocarbons to form ozonides these on hydrolysis give hydrogen peroxide and mixture of carbonyl compounds.

Ozonosphere. Layer in upper atmosphere, between 20 to 50 kilometres from surface of earth. It has appreciable ozone concentration, it absorbs sun's ultra-violet radiation and prevents high energy radiation from reaching earth's surface.

P

PABA (Para Amino Benzoic Acid). Component of folic acid which is necessary for formation of folic acid; present in many animals' and plants' tissue.

Pace Maker. Principal area in vertebrate heart from where there is discharge leading to depolarisation, resulting in contraction of heart. Small electric device inserted in human body for regulating pace of heart beat.

Pachy. Thick.

Packing Fraction. The difference the *relative atomic mass* of an *isotope* and its *mass number*, divided by the mass number.

PAH (Para Amino Hippuric Acid). Substance that is infused to measure plasma flow by determining urine and plasma concentration of substance.

Pain. Sense of discomfort on account of external/internal stimulation perceived through naked nerve endings.

Paint. Liquid that contains coloured material (pigment) in suspension.

Palisode Mesophyll. Main photosynthesising tissue of a leaf, situated below the upper epidermis and containing many chloroplasts.

Palmitic Acid ($C_{15}H_{32}COOH$). Hexadecanoic acid; wax like fatty acid, occurs as tripalmitin in palm oil and many natural fats; its salts form basis of soap.

Palmitin (Tripalmitin) [$C_{15}H_{31}COO)_3$. C_3H_5]. Glyceride of palmitic acid, boiling point 310 to 320°C., fat like substance which occurs in palm-oil and other fats and oils.

Palmitoyl. Univalent radicle $C_{15}H_{31}CO$—.

Panchromatic Film. A photographic *film* sensitive to *light* of all *colours* including red. It gives a more accurate representation of colours in monochrome than *orthochromatic film*.

Panthothenic Acid. A member of vitamin -B complex group and constituent of coenzyme- A; its deficiency leads to skin and alimentary canal disorders.

Pantoscope. Panoramic camera; wide-angled photographic lens.

Papain. An *enzyme*, present in the fruit and leaves of the paw-paw tree. It is capable of digesting *proteins*, used for softening meat for human consumption.

Papaverine ($C_{20}H_{21}NO_4$). A white *insoluble alkaloid*, m.p. 147°C., obtained from *opium*; used in medicine as an antispasmodic.

Paper. Sheets of *cellulose*, mainly obtained from wood pulp from which *lignin* and other non-cellulosic materials have been removed.

Paper Chromatography. A type of *chromatography* in which the mobile phase is liquid and the stationery phase is a strip of porous paper.

Para. Denoting positions at opposite apexes 'in a hexagonal ring of atoms, particularly the *benzene ring*. Abbreviated to *p*-as a prefix in naming a compound; *e.g.*, *p*-dichlorobenzene is 1, 4-dichloro-benzene.

Parabola. A curve traced by a point that moves so that its distance from a fixed point, the focus, is equal to its distance from a fixed straight line.

Parabolic Reflector. Paraboloid reflector. A *concave*, the section of which is a *parabola.*

Paraboloid of Revolution. The surface obtained by rotating a *parabola* around its *axis of symmetry.*

Paraffin (chem.). *alkanes.*

Paraffin Oil. Kerosine. A mixture of *hydrocarbons* obtained in the *distillation of petroleum.* The boiling range of the kerosines is 150°-300°C used for paraffin lamps, oil-burning engines, and domestic heaters.

Paraffin Wax. A white translucent *solid* that melts to a colourless *liquid* in the range 50°-60°C. It is a *mixture* of the higher *alkanes*, used for candles, waved paper and polishes.

Parallel Conductors. In Electrical *conductors* that are joined in parallel between two points *A* and *B*, so that each conductor joins *A* to *B*. If R_1, R_2, R_3, etc., are the resistances of the separate conductors, the total resistance *R* between *A* and *B* is given by the formula.

$$1/R = 1/R_1 + 1/R_2 + 1/R_3 \text{....etc.}$$

Parallel Beam of Light. A *beam* of *light* that neither converges nor diverges. *Lasers* are capable of producing nearly parallel beams.

Parallelepiped. A solid figure having six faces, all *parallelograms*: all opposite pairs of faces being identical and parallel.

Parallelogram. A plane four-sided *rectiliner* figure whose opposite sides are parallel.

Parallelogram of Forces. A particle under the action of two *forces* represented in direction and magnitude by the two sides of a *parallelogram* drawn from a point, the *resultant* is represented by the *diagonal* of the parallelogram drawn from that point.

Parallelogram of Velocities. A body having two component *velocities,* represented in magnitude and direction by two adjacent sides of *parallelogram* drawn from a point, the *resultant* velocity is represented by the *diagonal* of the parallelogram drawn from that point.

Paramagnetism. Magnetic properties caused due to unpaired electrons.

Parameter. A variable that can be kept constant while the effect of other variables is studied.

Parametric Amplifier. An *amplifier* of *microwaves,* it depends on the periodic *variation,* by an alternating *voltage,* of the *reactance* of a *thermionic valve* or *semiconductor* device.

Parasitic Capture. The absorption of a *neutron* by *a nuclide.* It does not result in a *nuclear fission* or the production of a useful artificial *element.*

Paris Green. A double *salt* of copper (II) ethanoate and arsenate (III), $Cu (CH_3COO)_2 . 3Cu(AsO_2)_2$. Used as a *pigment, insecticide,* and a wood preservative.

Parity. The principle of spacereflection symmetry, or conservation of parity, states that no fundamental distinction can be made between left and right ; that the laws of *physics* are the same in a right-handed system of coordinates as they are in a left handed system.

Parkes Process. Process of refining lead ; small quantity of zinc is added to molten impure lead ; silver being more soluble in zinc than lead, is then easily separated.

Parseer. An astronomical unit of distance, equal to 19×10^{12} miles, 3.26 *light-years,* or 3.084×10^{16} *metres.*

Partial Derivative. The *derivative* of a *function* with respect to only one of its *variables,* keeping all other variables in the function as *constant.*

Particle Physics. The branch of physics dealing with nuclear structure and the properties of *elementary particles.*

Partition Coefficient. The ratio of the concentrations of any *solute* in two immiscible *solvents,* at equilibrium.

Parton. A basic particle, like a *quark* that produces other *elementary particles.*

Parylene Polymers. A series of *polymers* obtained from di-1,4-xylene $(CH_2C_6H_4CH_2)_2$. Used for dielectric coatings in electronic equipment.

Pascal. The *SI Unit of pressure*, equal to 1 *newton* per square *metre.* Symbol Pa.

Pascal's Law of Fluid Pressure. *Pressure* of *fluid* causes a *force* to be transmitted aqually in all directions. This force acts at right angles to any surface within, or in contact with the fluid.

Paschen Series. A series of lines present in the *Infrared* region of the *spectrum* of hydrogen.

Paschen's Law. The *potential* required for sparking to occur, is a *function* only of the product of the *pressure* of the gas and the separation of the electrodes.

Passive. It denotes an electric component that does not amylify a signal.

Passive Absorption. Absorption of substances by a cell in accordance with the laws of diffusion.

Passivity. A state of *metals* in which it become resistant to *corrosion* on treatment with strong *oxidizing agents.* This is due to the formation of a surface *oxide* film.

Pasteurisation. It involves heating to a *temperature* sufficiently high to kill *bacteria.*

Pasteurism. Method given by Pasteur for inocculation with attenuated virus of certain diseases.

Patronuite VS_4. Ore of vanadium.

Pauli Exclusion Principle. No two electrons in an atom can have the same set of four quantum numbers.

Pearl. A secreation that consists mainly of *calcium carbonate*, $CaCO_3$, produces by various molluscs.

Pearl Ash. *Pottassium carbonate* K_2CO_3, obtained from wood *ashes.*

Pearl Essence. Silvery preparation from fish scales; used in making artificial pearls.

Pearlite. A microconstituent of iron or *steel.* It consists of alternate layers of *ferrite* and *cementite.*

Peat. Intermediate in the formation of *coal* from vegetable matter. It is an accumulation of partly decomposed plant material, used as *fuel.*

Peltier Effect. When an *electric current* is allowed to flow across the junction between two *metals* or *semiconductors*, a quantity of *heat*, proportional to the total *electric charge* crossing the junction, is evolved or absorbed, depending on the direction of the current.

Pencil Lead. A mixture of *graphite* and *clay* in various proportions, to give different degrees of hardness.

Pencil of Light (phys.). A collection of *rays* proceeding from or towards a point.

Pendulum, Simple. A device consisting of 'bob' swinging on the end of a string.

Penetration Factor. Refers to the probability of an incident particle, in a nuclear reaction, of passing through the nuclear barrier.

Panicillin. *Antibiotics* produced by the *Penicillium* mould. A very powerful agent to prevent the growth of several types of disease *bacteria.*

Pennyweight. 24 grains, 1/20 Troy ounce.

Penta. Prefix denoting five.

Pentachlorophenol. C_6CI_5OH. A white *insoluble derivative* of *phenol,* m.p. 174°C., used as a *fungicide.*

Pentaerythritol. $C(CH_2OH)_4$. A white *soluble* powder, m.p. 260°C., used in the manufacture of *plastics,* plasticizers, and *explosives.*

Pentagon. A five-sided *polygon.*

Pentane C_5H_{12}. It is contained in light *petroleum; n*-pentane has b.p. 36°C., and r.d. 0.62, used as a *solvent.*

Pentanoic Acid. Valeric acid. C_4H_9COOH. A fatty acid, a colourless *liquid,* b.p. 186°C., having a pungent odour, used in perfumes.

Pentanol. Amyl alcohol. $C_5H_{11}OH$. A colourless *liquid,* b.p. 137.3°C., having characteristic smell.

Pentavalent. Quinquevalent. Having a *valence* of five.

Pentlandite. A bronze sulphide *mineral,* $(Fe, Ni)_9\ S_8$, the chief *ore* of nickel.

Pentode. A *thermionic valve* having five *electrodes*: a *cathode,* an *anode* or plate, a *control grid,* two other grids known as *screen grid* and the *suppressor grid.*

Pentosans. *Polysaccharide* yielding *pentoses* on *hydrolysis.*

Pentose. A *monosaccharide* containing five carbon *atoms.* The most important is *ribose* an essential constituent of the *nucleic acids.*

Pentyl. The *univalent radical* C_5H_{11}—, formerly called the *n-amyl* radical.

Pentyl Ethanoate. Amyl acetate, banana oil. $CH_3COOC_5H_{11}$. A colourless liquid, b.p. 148°C., with an odour of pear drops, used as a solvent and in perfumes and flavours.

Penumbra. Half shadow it is formed when an object in the path of ***rays*** from a large source of *light* cuts off a portion of the light.

Pepsin. Enzyme related with protein digestion in animals.

Pepsinogen. Inactive form of pepsin as it is secreted, to prevent autodigestion of body in the individual; it is activated by acidity of food in stomach.

Peptidase. An *enzyme* which attacks *peptide* linkages and splits off *amino acids.*

Peptide. A *compound* of two or more *amino acids* formed on ***condensation*** of the NH_2 group of one acid and the *carboxyl group* of another. The peptide linkage, —NH —CO.

Peptones. Organic substances produced on *hydrolysis* of ***proteins*** by the action of *pepsin* in the stomach.

Per. Prefix denoting, an excess of the normal amount of an *element* in a *compound*; *e.g., peroxide.*

Perborate. A *salt* of *perboric acid*, HBO_3.

Perboric Acid HBO_3. A hypothetical *Acid* known only in the form of its *salts*, the *perborates*; *e.g., sodium perborate.*

Perching. Having a firm grip on objects (usually stems and branches or even cables) with toes and digits by birds.

Perchlorate. A *salt* of $HClO_4$.

Perchloric Acid. $HClO_4$.

Percussion Cap. A device used in fire-arms. It consists of a small copper cylinder that contains *mercury cyanate* or other violent *explosive* which explodes on being struck, initiating the explosion of the main charge.

Perfect Gas. Ideal gas. Such a gas consists of perfectly elastic ***molecules***, the *volume* occupied by the actual molecules, and the *forces* of attraction between them, being zero or negligible. It obeys gas laws.

Pericardinum. Refers to them membrous sac found around heart in vertebrates.

Periclase. MgO, ore of Mg.

Pericynthion. Refers to the time of, or the point of, nearest approach of a satellite in the lunar orbit to the moon's surface.

Perigastritis. It refers to the inflamation of outer surface of stomach.

Perigee. The sun, the moon or an artificial earth satellite are said to be in perigee when they are at their least distance from the earth.

Perigynous. Flowers having margin of thalamus upward. They form cup-shaped structure enclosing ovary but free from it. It carries with it, sepals, petals and stamens.

Perihelion. The time of or point of nearest approach of a planet to the sun.

Perimeter. The distance around a plane figure.

Periodate. A *salt* of *iodic (VII) acid* (periodic acid). (HIO_4).

Periodic Acid. See *iodic (VII) acid* (HIO_4).

Periodic Law. 'The properties of the element are periodic function of their *atomic weights.*'

Periodic Table. An arrangement of the chemical *elements* in order of their *atomic numbers* that demonstrate the *periodic law.* In such an arrangement. Elements with the same numbers of electrons in their outer shells fall into the same vertical group and have similar chemical properties.

Peripherals. Devices connected to the *C.P.U.*, or the high-speed *store* of a *computer.*

Periphery. Boundry of a body; the circumference of *perimeter* of any closed figure.

Periscope. Optical instrument; long tube equipped with lenses and mirrors, arranged in such a way that person looking through one end (lower) can see objects reflected at other (upper) end; used in submarines, trenches etc.

Permalloy. Iron-nickel *alloys* with high *magnetic permeability* used in parts of elctrical machinery that are subject to alternating *megnetic fields*, also used in *computer* memories.

Permanent Gas. A *gas* that cannot be liquefied by *pressure* alone; a gas above its *critical temperature.*

Permanent Hardness of Water. Hardness that cannot be destroyed by *boiling* the *water.*

Permeability. Ease of flow of a substance through a membrane.

Permease. Enzyme that is responsible for translocation of ion(s) across a membrane.

Peroneus. It is one of several fibular muscles.

Peroneus Longus. Superficial flexor muscle found in leg of perching birds.

Peroxidase. Enzyme which is capable of breaking peroxide present in white blood corpuscles like neutrophils.

Peroxide. Oxide with highest proportion of oxygen (O_2); this yields hydrogen peroxide on treatment with acid.

Peroxisomes. Microbody containing amino acid oxidases, urate oxidases and catalase.

Perpetual Motion. The concept of a machine that, once set in motion will go on for ever without receiving *energy*.

Persistence of Vision. The sensation of *light*, persists for a brief interval after the actual light stimulus is removed.

Personal Equation. The time interval peculiar to a person between the perception and recording of any event.

Perturbations. Deviations in the motions of the *planets* from their path due to their gravitational attractions for each other.

Pest. Any organism (plant or animal) causing loss or annoyance to man or his property to such an extent that it warrants control.

Pesticide. Chemical used to kill the pest; it may be acaricide, bactericide, fungicide, herbicide, insecticide, molluscide, nematicide, ovicide, rodenticide, viricide, weedicide etc.

Pestiferous. That brings plague or pestilene.

Pestilene. Any deadly epidemic disease.

Pestology. Study of agricultural pests and their control.

Peta (P). It denotes one thousand million million, 1 *pm* = 10^{15} metres.

Petrification. Change of an organic structure into a mineral structure.

Petrochemicals. Chemicals obtained from petroleum or natural gas.

Petrol. Gasoline; light volatile complex mixture mainly of hydrocarbons such as heptane, hexane and octane; obtained by fractional distillation or cracking of petroleum; used to drive various engines including those of cars and aeroplanes.

Petrolatum. Petroleum jelly; mixture of semi solid hydrocarbons yellowish soft-paraffin, often misnamed as vaseline.

Petroleous. Containing petrol.

Petroleum. Naturally occurring complex mixture of paraffin, cycloparaffinic and aromatic hydrocarbons having low percentage of sulphur and trace amounts of nitrogen and oxygen compounds; it yields combustible fuels , petro-chemicals and lubricants.

Petroleum Ether. Mixture of lower hydrocarbons of alkanes series mainly of pentane and hexane.

Petrology. Relating to the study of origin, structure and composition of rocks.

Pewter. Alloy of three to nine parts tin (Sn) and one part lead (Pb) with little of antimony to harden it, or copper to soften it.

Pfitzer's Law. Cells progressively diminish during successive generations.

PH. *Pouvoir Hydrogene* or hydrogen power; indicating relative concentration of hydrogen ions in solution.

PH Meter. Instrument used to find out pH of solution.

Phagocytosis. Engulfing of food particles due to membrane invagination proximal to particle and subsequent membrane closure over the particle.

Pharmacology. Study of action of chemical substances on animals, science of drugs.

Pharmacy. Place for the preparation and dispensing of drugs.

Phase (chem.). A separate part of a *heterogeneous* body or system.

Phase Angle. 1. (phys.). The *angle* between the vectors representing two harmonically varying quantities having the same *frequency.* **2.** (astr.) The *angle*, seen from the *Moon* or a *planet*, between the *Earth* and the *Sun.*

Phase Contrast Microscope. A *microscope* that uses the difference in *phase* of the *light* transmitted or reflected by an object to form an *image* by relative differences in *intensity.*

Phase Diagram. A diagram that shows the relations between various *phases* in a chemical system, and the effects of composition and conditions on them.

Phase Modulations. *Modulation* of the *phase angle* of a *sinusoidal carrier wave.* The *phase* of any modulated wave differs from that of the carrier by an amount which is proportional to the instantaneous value of the modulating wave.

Phase Rule. $F+P = C + 2$. For a *heterogeneous* system in equilibrium. The sum of the number of *phases* plus the number of *degrees of freedom* is equal to the number of *components*, plus two.

Phases of the Moon. The various shapes of the illuminated surface of the *Moon* as seen from the *Earth.*

Phasor. A rotating *vector* that represents a quantity varying *sinusoidally.* Its length represents the *amplitude* of the quantity.

Pheasant. A rich coloured chicken like bird. It is about 80-90 cm long, has long tail and brilliant feathers. Its flesh is used as food.

Phenacetin. 1, 4-ethoxyphenylethanamide. $CH_3CONHC_6H_4OC_2H_5$. A white crystalline substance, m.p. 134.7°C., used as pain releiver and as an *antipyretic.*

Phenazine. $C_6H_4N_2C_6H_4$. A yellow crystalline substance, m.p. 171°C., used in the manufacture of *dyes.*

Phenetole ($C_6H_5OC_2H_5$). Ethoxybenzene. A *volatile liquid*, b.p. 172°C.

Phenobarbitone. Phenylethylbarbituric acid. $C_6H_5C_2H_5.C:(NHCO)_2:CO$. A white crystalline powder, m.p. 174°C., used as a sedative and *hypnotic.*

Phon. Unit of loudness used to measure intensity of sound.

Phoneme. One of the basic sounds on which speech is made up.

Phonograph. Instrument used for reproducing sound—gramophone.

Phonon. Quantum of thermal energy in lattice vibrations of a crystal.

Phosgene. Carbonyl chloride. $COCl_2$. A colourless poisonous *gas* with a penetrating smell. It was used in World War I as a war *gas.*

Phosphamidon ($C_{10}H_{19}O_5NCIP$). Organophosphate, pale yellow oil miscible with water readily soluble in most organic solvents used as systemic acaricide and insecticide.

Phosphatase. Enzyme which splits phosphate radicals from their esters.

Phosphate. Salt of phosphoric acid.

Phosphaturia. Refers to excess of phosphates in urine.

Phosphide. Compound of phosphorous with another element.

Phosphine (PH_3). Highly poisonous, colourless gas obtained on treating phosphides with water; used in smoke screen and as Holme's signal for ships.

Phosphinic Acid (H_3PO_2). Colourless crystalline solid' obtained by heating white phosphorus with an alkali solution. Soluble in water; some of its salts are used as nerve tonics.

Phosphite. Salt of phosphorous acid.

Phosphodiester. Molecule containing two alcohols esterified to one molecule of phosphoric acid.

Phospholipids. Compound lipids containing phosphoric acid groups and nitrogenous base: found in brain tissue and egg yolk.

Phosphonic Acid (H_3PO_3). Phosphorus acid, white deliquescent diabasic acid (solid), obtained on dissolving phosphorus trioxide in cold water, soluble in water.

Phosphonium (PH_4^+). Radical analogous to ammonium.

Phosphor. A substance capable of *luminescence.*

Phosphorate. To combine with phosphorous.

Phosphor-bronze. Alloy of copper (80 to 95%), tin (4 to 18%) and phosphorus (0.25 to 2.5%); hard, tough and elastic.

Phosphoresce. To undergo phosphorescence.

Phosphororescence. Property of emitting off light after exposure to light, X-rays etc.

Phosphoric Acid (H_3PO_4). Colourless, syrupy liquid, weak tribasic acid obtained by treating phosphorite rock with dilute suphuric acid.

Phosphorism. Refers to poisoning by phosphorus.

Phosphorite. Impure massive apatite.

Phosphogrogen. Substance which promotes phosphorescence in another substance, *e.g.*, manganese does in zinc sulphide.

Phosphorolysis. Reaction by which elements of phosphoric acid are introduced into molecule of compound.

Phosphorous Acid (H_3PO_3). Colourless deliquescent crystalline substance.

Phosphorus (P). Non-metallic element, occurs in two main allotropic forms, boiling point 280 to 281°C—white phosphorus is waxy white solid, very inflammable and poisonous; red phosphorus is non-poisonous, dark red powder, not very inflammable ; both show the property to give out light in the dark.

Phosphorous Pentabromide (PBr_5). Yellow solid boiling at 106°C, obtained by action of bromine on phosphorus tribromide; hydrolysed by water.

Phosphorous Pentachloride (PC_{l5}). Pale yellow solid obtained by action of excess of chlorine or phosphorous trichloride; fumes in air and has pungent odour, reacts violently with water and other organic compounds.

Phosphorus Pentaoxide (P_2O_5). White solid, it forms phosphoric acid (H_3PO_4) on treatment with excess of water.

Phosphorus Tribromide (PBr_3). Colourless liquid obtained by adding phosphorus to solution of bromine in benzene, hydrolysed by water into phosphorus acid and hyrobromic acid.

Phosphorus Trichloride (PCl_3). Mobile liquid boiling at 7+°C., obtained by action to dry chlorine on yellow phosphorus; fumes in moist air and reacts violently with water, making hissing sound and yielding phosphorus acid, $PCl_3+H_2O \rightarrow 3H_3\text{-}PO_3+3HCl$; used in organic reactions and in preparation of phosphorus pentachloride.

Phosphorous Triodide (P_4O_6). White solid with garlic-smell, boiling point 173°C, readily soluble in water to form acid (H_3PO_3); used in preparation of phosphorous acid and phosphites.

Phosphoryl. Trivalent radical ≡ PO.

—

Phot. Unit of illumination ; one lumen per square centimeter.

Photic. Which is sensitive to light.

Photics. Optics.

Photoautotroph. Autotroph that utilises energy of solar radiation to synthesise organic food from inorganic substances.

Photobiology. Branch of biology dealing with effect of light on organism.

Photocathode. Cathode that emits electrons on being illuminated; result of photoelectric effect.

Photochemistry. Study of such chemical process as are initiated by light or electromagnetic radiation.

Photochromism. Property of certain materials to undergo reversible change in colours on being exposed to light of different wavelengths.

Photoconduction. Increase in electrical conduction arising from absorption of electromagnetic radiation.

Photoconductivity. Property of changing conductivity under influence of light.

Photocap. It is a photographic copy of printed or other graphic material.

Photodiode. Semiconductor diode in which reverse current varies with illumination.

Photodisintegration. Nuclear reaction caused by a photon in which the nucleus emits charged fragments of neurons.

Photodissociation. Dissociation of chemical compound as a result of absorption of radiant energy.

Photoelasticity. When certain materials (*e.g., glass, perspex*, etc.) are stressed they become doubly refracting.

Photoelectric Cell. Photocell. A device used for the detection and measurement of *light* and other radiations.

Photoelectric Effect. Any effect arising out of a transfer of *energy* from *light* incident on a substance.

Photoelectricity. Electricity produced by light or some other electromagnetic radiation.

Photoelectron. Electron ejected from a body when ultra-violet or X-rays are incident upon it, or by photoionisation.

Photoelectronics. Science that deals with interaction of electricity and light.

Photoelectron Spectroscopy. A method of determining *ionization potentials* and molecular structure.

Photoemissive. Capable of emitting *electrons* on being subjected to *electromagnetic radiation.*

Photoemissive Cell. Device that is used to detect or measure radiant energy by measurement of resulting emission of electrons from surface of photocathode.

Photoemissivity. Property of substance that emits electrons on being struck by light.

Photofission. Refers to nuclear fission by photon.

Photogen. Light producting organ in animals.

Photoglhyph. Photographic engraving; photogravure.

Photograph. Picture taken with the help of camera.

Photographic Camera. Equipment used to take photographs—still or moving.

Photography. The complete process of taking an image of the object on a photographic paper.

Photogravure. Method of photo-engraving using acid on sensitised material.

Photoionisation. Removal of one or more electrons from an atom or molecule on exposure to radiation.

Photology. Scientific study of light.

Photoluminescene. *Luminescence* caused by *electromagnetic radiation.*

Photolysis. Decomposition under influence of radiation. Grouping of chloroplasts in relation to illumination.

Photomeson. Meson produced due to interaction between a photon and an atom nucleus.

Photometer. Apparatus for comparing illuminating power of two sources of light.

Photometry. Branch of physics dealing with measurement of light energy emitted or absorbed by a body.

Photomicrograph. Enlarged photograph of microscopic objects taken through microscope.

Photomorphosis. Change in structure of plant on being exposed to strong light.

Photon. Quantum of light.

Photonasty. Response given by some plants to stimuli of changed intensity of light without regard to direction.

Photoperiod. Optimum length of one day ; it affects amount of light received by a plant, for its normal growth and development.

Photoperiodism. Response of plant to relative lengths of day and night.

Photophil. Light-loving.

Photophile. Phase in which flowering is promoted by light.

Photophobe. Phase in which flowering is inhibited by light.

Photophobia. Shrinking from light.

Photophone. Apparatus used for transmitting articulate speech to a distance along a beam of light.

Photophore(s). Luminiferous organ; modified mucous glands of deep-sea-fishes. They produce light to warn enemies and attract prey.

Photophoresis. Migration of suspended particles under influence of light.

Photopic Vision. Vision in which the cones in the eye are the principle receptors.

Photopigment. Pigment molecules that can be excited by light.

Photo-process. Process of printing using photographically prepared plates.

Photoproton. Proton released from nucleus in photonuclear reactions.

Photopsia. Appearance of flashes of light due to irritation of retina.

Photoreceptor. Light sensitive spot; eye of vertebrate.

Photoreduction. Light induction of an electron acceptor in photosynthetic cells.

Photosensitive. Sensitive to radiant energy ; affected by visible or invisible light.

Photosphere. Intensely luminous and visible portion of sun ; has temperature around 6000°K.

Photostat. Apparatus used for making facsimiles of any written or printed material, drawings etc., directly as positives.

Photosynthesis. Process by which green plants use solar energy and convert the same into potential chemical energy.

Photosynthetic. Process by which green plants use solar energy and convert the same into potential chemical energy.

Photosynthetic Bacteria. Group of bacteria that are able to photosynthesise through possession of a pigment (bacteriochlorophyll).

Phototaxis. Change of place under light stimuli.

Phototelegraph. Apparatus used for transmitting drawings, photographs etc., by telegraphy.

Phototheraphy. Treatment of disease using light.

Phototroph. Organism capable of deriving energy necessary for synthesis of organic compounds directly from solar radiation.

Phototropism. Orientation in response to light stimuli. Reversible colour change on exposure to light.

Phototype. Printing block that is produced photographically.

Photovoltaic Effect. A *photoelectric effect* in which *light* falling on a boundary between certain pairs of substances (*e.g.*, copper and *copper (I) oxide*) produces a *potential difference* across the boundary.

Photo-xylography. Wood carving after taking photographic impression on wood block.

Photo-zincorgraphy. Process of engraving on zinc, by taking first photographic impression on zinc plate then etching with acids.

Phthalic Acids. $C_6H_4(COOH)_2$. Three *isomeric* acids. 1. Phthalic acid, the *ortho* form, 1,2-benzenedicarboxylic acid, is a white crystalline solid, m.p. 207°C., that decomposes into *phthalic anhydride* and water and is used in organic synthesis 2. The *meta* form, 1, 3-benzenedicarboxylic acid. 3. The *para* form, 1,4-benzenedicarboxylic acid.

Phthalic Anhydride. $C_6H_4(CO)_2O$. The *anhydride* of *o-phthalic acid*, can be obtained by *oxidation* of *naphthalene* in the presence of acatalyst.

Phthalocyanines. Organic colouring matters, usually of outstanding resistance to the action on light and other agencies.

Physical Change. Any change in a body or substance not involving an alteration in its chemical composition.

Physical Chemistry. The study of the *physical changes* associated with *chemical reactions* and the dependence of physical properties on chemical composition.

Physical States of Matter. The physical state in which *matter* exists, at a particular *temperature* and *pressure.* In *gases* molecules are free to move about the space, that contains them, almost independently of each other. In *solids* they vibrate about the fixed poisitions of a *crystal lattice.*

Physics. The study of the properties of *matter* and *energy* .

Physiological Saline. An *isotonic solution* of *salts* in *distilled water,* used for preserving cells.

Physiology. The study of the functioning of the various parts of living beings.

Physiotherapy. Therapy that uses exercise, massage, heat etc.

Physique. Refers to structure, constitution, strength or appearance of body.

Pi (π). Symbol for ratio of circumference of circle to its diameter (=3.14159 or $\frac{22}{7}$).

Pi Bond. π bond.

Pico. Prefix denoting one millionth, *e.g.*, a picefarad is 10^{-12} *farad.*

Picoline. Methylpyridine. $CH_3C_6H_4N$. A heterocyclic compound exists in three *isomeric* forms. All isomers are found in *coal-tar* and *bone oil*; they are used as *solvents* and as *intermediates* in *organic synthesis.*

Picrate. A *salt* of *picric acid.*

Picric Acid. 2,4,6-trinitrophenol. $C_6H_2(NO_2)_3OH$. A bright yellow crystalline *solid,* m.p. 122°C, used as an explosive, as a *dye* and (in *solution*) for treating burns.

Pie Chart. A diagrammatic chart representing percentage as slices of a circular pie, *i.e.*, sectors of a *circle.*

Piezoelectric Effect. A property of certain *asymmetric crystals*, such as *Rochelle salt* or *quartz.*

Pigeon's Milk. Slim, fatty, proteinous substance secreted by crop glands in both sexes of pigeons, regurgitated by parents into their beaks and then fed into mouth of nestlings, to nourish them.

Pig-Iron. An impure form of iron.

Pigment Colour. Body colour. The *colour* of most natural objects is due to the absorption by the substance of the different *wavelengths* present in the incident white *light.* The colour the body appears is determined by the wavelengths absorbed the least.

Pigments. 1. Materials used for imparting various colours to *paints, plastics,* etc. **2.** Natural colouring substances in plant or animal tissues.

Pilocarpine ($C_{11}H_{16}N_2O_2$). White crystalline alkaloid, used in medicines,. (MP = 34°C).

Pinchbeck. An *alloy* of copper and zinc, used as an imitation gold.

Pinch Effect. 1. The construction of a *liquid conductor* of *electricity* (*e.g.*, mercury on passing substantial *current* through it). **2.** The construction of a *plasma* owing to the *magnetic field* of a high current within the plasma.

Pink Salt. $(NH_4)_2SnCl_6$, used as a mordant in dyeing.

Pint. A unit of capacity equal to one eighth of a *gallon.*

Pion. A pi-*meson.* A type of meson.

Piperazine. Hexahydropyrazine. $C_4H_8(NH)_2$. A colourless *deliquescent* solid m.p. 108-110°C., used mainly as a *vermifuge.*

Piperidine ($C_5H_{10}NH$). A colourless *liquid,* b.p. 106°C., used as a solvent.

Piperine ($C_{17}H_{19}NO_3$). A white crystalline *alkaloid,* m.p. 129.5°C., constituent of pepper.

Pipette. A glass tube with which a definite *volume* of *liquid* may be transferred.

Pirani Gauge. A pressure gauge used to measure low pressures.

Pitch. Residue left in destructive distillation of wood, coal tar etc.

Pitchblende. Uraninite; natural ore containing mainly uranium oxide (U_3O_8); it is principal source of radium.

Pilot Tube. An instrument for measuring the speed of a *fluid*; consisting of a tube that has two openings, one facing the moving fluid and the other facing away from it. The difference in *pressure* created in the tube between the two openings, as measured by a *manometer,* allows the speed of the fluid to be determined.

PK. Defined as log 1/*K*, where *K* is the *equilibrium constant* of the *dissociation* of the acid. The higher the value of pK, the weaker is the acid.

Planck Constant *h.* The constant relating the *frequency* of a *radiation,* ν, with its *quantum* of *energy*, *E*; *i.e.*, $E=h\nu$.

Planck's Law of Radiation. The energy radiated by unit surface area of a black body at a thermodynamic temperature, *T*, in unit time is:

$$2\pi hc^2\lambda^{-5}/(\text{exp.}\ \lambda\ Thc/k\ -1)$$

where *h* is the *Planck constant*, *c* is the *speed of light*, *k* is the *Boltzmann constant*, and λ is the wavelength of the radiation.

Plane (math.). It is defined as surface containing all the straight lines passing through a fixed point and also intersecting a straight line in space.

Planetarium. 1. A complex system of optical projectors to represent the movements of the *planets* and *stars* on a domed ceiling. **2.** The build housing such a system.

Planets. Heavenly bodies revolving in definite *orbits* about the *Sun.* They are *Mercury, Venus,* the *Earth, Mars, Jupeiter, Saturn, Uranus*

Neptune and *Pluto.*

Planimeter. A mechanical integrating instrument used for measuring *plane* areas, consisting of a movable tracing arm the movements of which are recorded on a dial.

Plaque. A deposit of dissolved food *saliva* and *bacteria* on the exposed enamel surfaces of teeth.

Plaster of Paris ($CaSO_4 . \frac{1}{2} H_2O$). Powdered calcium sulphate, obtained on heating gypsum to 120 to 130°C, when mixed with water it sets and hardens.

Plasticizer. 1. A *non-volatile liquid* added to *paints* and varnishes in order to avoid brittleness of the dried film. **2.** A liquid or *solid* substance added to synthetic or natural *resins* to improve their flow properties.

Plastics. Materials that are *stable* in normal use, but at some stage of their manufacture are plastic and can be shaped or moulded by *heat, pressure*, or both.

Platelets. Tiny granulated bodies 2 to 4 μm in diameter, millions of these are found in circulating blood; formed by mega-karyoblasts of red bone marrow; these help in blood coagulation.

Plate Tectonics. The theory that the *Earth's* crust consists of a number of semirigid plates, which move relative to each other.

Platinic. Containing platinum in its + 4 *oxidation state e.g.*, platinum (IV) chloride, $PtCl_4$.

Platiniferous. Platinum-bearing.

Platinized Asbestos. *Asbestos* having a black deposit of finely divided platinum used as a *catalyst.*

Platinoid. An *alloy* of 60% copper, 24% zinc, 14% nickel, and 2% tungsten.

Platinous. Containing platinum in its + 2 *oxidation state, e.g.*, platinum (II) chloride, $PtCl_2$.

Platinum. Pt. Element, R.a.m. 195.09. At. No. 78. A hard silvery white ductile and malleable metal, r.d. 21.45, m.p. 1772°C., b.p. 3827°C. It occurs as the metal, alloyed with osmium, iridium, and similar metals. It is used for electrical contacts, scientific apparatus, as a *catalyst* and in jewellery.

Platinum Black. Velvety black powder obtained when platinum is precipitated from solution of its tetrachloride by reducing agent like sodium formate, glucose etc., used as catalyst.

Platinum (II) Chloride ($PtCl_2$). Brown powder obtained by heating chloroplatinic acid to 580°C., or by heating spongy platinum in current of chlorine, insoluble in water but soluble in hydrochloric acid.

Platinum (IV) Chloride ($PtCl_4$). Brown solid formed on heating chloroplatinic acid to 350°C., in chlorine, $H_2PtCl_6 \rightarrow 2HCl + PtCl_4$.

Platinum Lamp. Electric lamp having platinum filament.

Platinum Metals. A group of *six transition elements i.e.*, ruthenium rhodium, palladium; osmium, iridium and platinum.

Pleochroic. Denoting certain crystals having different colours, depending on the direction from which they are observed.

Plumbago. Black-lead, *graphite*, natural *form of* carbon.

Pluto. A *planet* with its *orbit* outside that of *Neptune.*

Plutonium. Pu. At. No. 94. *Transuranic element.* Thirteen different *isotopes* of plutonium can be produced by suitable *nuclear reactions.* The isotope plutonium-239 is produced in *nuclear reactors* has a *half- life* of 24400 years, is used in *nuclear weapons*, one *kilogram* having an *energy* equivalent of about 10^{14} *joules.*

Pneumatic. Operated by, or filled with, compressed air.

Pnicogens. Used for the V group elements *i.e.* N, P, As, Sb and Bi.

Polyamide. A *polymer* in which the units are linked by *amide* or thioamide groupings.

Polybasic Acid. *Acid* containing more than one *atom* of *acidic hydrogen* in a *molecule.*

Polycarbonate. These thermoplastic resins in which structural units are linked through carbonate radicals.

Polychromatic. Having many colours.

Polycyclic. Having several rings, whorls or circuits.

Polyene. Organic compound having more than two double bonds:

Polyester. A *polymer* formed from a *polydydric alcohol* and a *polybasic acid* used in the manufacture of synthetic *resins, fibres* and *plastics.*

Polyethene. *Polythene.*

Polyethylene. *Polythene.*

Polygon. A *plane* figure bounded by straight line.

Polygon of Forces. If any number of *forces*, acting on a particle, can be represented in magnitude and direction by the sides of a *polygon* taken in order, then the forces will be in *equilibrium.*

Polygynous. Mating with several females.

Polyhedron. A solid figure having *polygons* for its faces.

Polyhybrid. Cross between parents which differ in many heritable characters.

Polyhypric. Compound containing more than one hydroxyl group *e.g.*, glycol and glycerol.

Polymer. Compound obtained by polymerisation having same empirical formula but greater molecular weight.

Polymerase. Enzyme which catalyses a biological polymerisation reaction.

Polymeric. Compound of same elements in same proportions by weight but differing in molecular weight.

Polymerisation. Union of molecules of same compound to form large molecules (polymer).

Polymethyl Methacrylate. Poly (methyl 2-methylpropenoate), Perspex. A colourless *solid thermoplastic* obtained by the *polymerization* of methyl methacrylate which is widely used in place of *glass.*

Polymorphism. The existence of the same substance in more than two different crystalline forms.

Polynomial (math.). An expression consisting of three or more terms.

Polynucleotide. A chain of *nucleotides* linked together as in a *nucleic acid. Ribonucleic acid* consists of a single chain, while *deoxyribonucleic acid* usually consists of a double *helix* comprising two polynucleotide chains.

Polyonymous. Having many names.

Polypeptide. Chain consisting of three or more amino acids each of which is joined to its neighbour by peptide linkage.

Polypetalous. Flower having its various petals free from each other.

Polyploid. Having more than twice the normal haploid number of chromosomes.

Polypod. Animal having many feet.

Polypodiaceae. Family of ferns having stalked sporangia and vertical annulus.

Polybody. Any fern that belongs to genus *Polypodium.*

Polypropene. Polypropylene. A *transparent thermoplastic* material obtained by the *polymerization* of *propane.* Used as flexible *plastic* similar to *polythene* but possess more strength.

Polysaccharides. Natural carbohydrates obtained from condensation of several molecules of simple sugars (monosaccharides)—cellulose, starch.

Polysepalous. Calyx with its sepals freely (separate from each other inserted on receptacle).

Polystyrene. A *thermoplastic* material, obtained by the *polymerization* of *styrene* (phenylethene ; $C_6H_5CH : CH_2$). It is a clear glassy material possessing good insulating properties.

Polytene. That have many duplicate chromatin strands.

Polytetrafluoroethane. PTFE. Teflon. Fluon. A *thermosetting* material obtained by the *polymerization* of *tetrafluoroethene* ($CF_2 : CF_2$).

Polythene. Polyethene, polyethylene, alkathene. A tough *thermoplastic* obtained by the addition *polymerization* of *ethene*, C_2H_4. Used as an insulting material.

Polyvinyl Acetate. PVA. Polyethenyl ethanoate. A *thermoplastic* obained by the *polymerization* of *vinyl acetate* (ethenyl ethanoate; $CH_2 : CHOO (CH_3)$. Used in *adhesives, inks*, and lacquers for coating paper and fabric.

Polyvinyl Chloride. PVC. Polychloroethane. A *thermoplastic* obained by the *polymerization* of *vinyl chloride* (chloroethene; $CH_2:CHCl$). It is resistant to *water, acids, alkalis*, and *alcohols*.

Polyvinylidene Chloride. Polydichloroethene. A *thermoplastic* obtained by the *polymerization* of *vinylidene chloride* (dichloroethene $CH_2 : CCl_2$). Also used as a copolymer with *acrylonitrile* or *vinyl chloride* to give products having a wide range of flexibilities.

Polywater. A form of water, differing in properties (*density, viscosity*) from normal water, that was reported in 1962. It is now accepted that these properties were due to the presence of colloidal particles derived from impurities rather than to any difference in the molecular structure of the water itself.

Population Inversion. The existing situation in a *laser* when a large proportion of the emitting *ions* have been raised to an excited state by the process of optical pumping.

Population Type. Refers to a classification of *stars* into two types: Population I includes hot white young stars such as those forming the spiral arms of *spiral galaxies*; Popuation II consists of older stars, such as *red giants* which are found at the centres of spiral galaxies.

Porcelain. A hard white material obtained by the firing of a *mixture* of pure kaolin (*china clay*) with *felspar* and *quartz*, or with other materials containing *silica.*

Pore. Small passage or interstice, opening of sweat gland.

Porphyrins. A class of naturally occuring *pigments* obtaind from pyrrole. Such as *chlorophyll* and the haem of *haemoglobin.*

Position Circle. A *circle* having its centre at an observed point and a circumference passing through the place of observation.

Position Line. A line where an observer is situated at a given time. The intersection of two position lines, determined at the same time, fixes the observer.

Positive (math., phys). In any convention of signs, as opposed to negative.

Positive Column. A luminous region in a *discharge in gases* near to the positive *electrode.*

Positive Lens. Lens that has positive (+) focal length or power—convex lens.

Positive Magnetic Pole. The north-seeking pole of a *magnet.*

Positive Rays. Streams of *ions* carrying positive *electric charges*. They are produced by an electric discharge in a rarefied gas. See *discharge in gases.*

Positron. Anti-particle having mass and spin same as of electron; produced during pair production and during several decay processes.

Positronium. Unstable assembly of positron and electron ; it decays into protons.

Post Caval. Large vein draining posterior body region and that opens into right auricle.

Posterior. Coming after, behind, hind or rear.

Potash. Old name for K_2CO_3, KOH or any potassium salt.

Potassium (K). Silvery-white soft metal prepared by electrolysis of potassium cyanide or potassium hydroxide or potassium fluoride and calcium carbide, boiling point 756°C ; essential to life and found in all living matter; used in making alloys and in photo-electric cells.

Potassium-argon Dating. A method of *dating* geological specimens that is based on the decay of the *radioisotope* potassium-40 to argon-40. The *half-life* of potassium-40 is about 1.3×10^9 years.

Potassium Bicarbonate ($KHCO_3$). White solid prepared by passing carbon dioxide through cold solution of potassium carbonate; soluble in water and forms alkaline solution.

Potassium Bromide (KBr). White crystalline solid prepared by saturating hot solution of caustic potash with bromine, the solution is dried and again heated with charcoal to get potassium bromide; used medically as sedative and in photography.

Potassium Carbonate (K_2CO_3). Pearl ash ; white hygroscopic solid, highly soluble in water produces alkaline solution.

Potassium Chlorate ($KClO_3$). Fine white flakes, obtained on electrolyses of sodium chloride solution at 40°C and treating the solution so obtained with potassium chloride used as oxidising agent and in making matches, fireworks etc.

Potassium Cyanide (KCN). White crystalline solid; deadly poisonous, highly hygroscopic and very soluble in water; used for extraction of gold and silver.

Potassium Hydroxide (KOH). Caustic potash, white highly deliquescent solid obtained by electrolysis of solution of potassium chloride; used in soft toilet soaps etc.

Potassium Iodide (KI). White crystals, soluble in alcohol and water, obtained by adding iodine to hot concentrated solution of caustic potash which is dried and residue heated with charcoal to get potassium iodide ; used mainly in photography.

Potassium Permaganate ($KMnO_4$). Dark violet, anhydrous needlelike crystals having metallic lustre, obtained by electrolytic oxidation of magnate ; used as oxidising agent, disinfectant and as Baeyer's reagent in organic chemistry.

Potassium Nitrate (KNO_3). Saltpetre; white needle-like crystals, soluble in water used in manufacture of fire-works, gun-powder nitric acid etc.

Potassium Sodium Tartrate. White crystalline soluble salt ; used in preparing baking powder.

Potassium Sulphate (K_2SO_4). Colourless rhombic crystals obtained by action of sulphuric acid on potassium chloride ; used in making alum, hard glass and fertilizers.

Potassium Thiocyanate (KSCN). Colourless hygroscopic substance; used in manufacture of drugs and dyes.

Potential Difference. It is defined as the *work* done when a unit positive *electric charge* is moved from one of the points to the other. The practical unit of p.d., is the *volt.*

Potential Energy. The *energy* possessed by a body by virtue of its position *e.g.*, a coiled spring, or a vehicle at the top of a hill. The potential energy of a *mass, m,* raised through a height, *h*, is ***mgh*,**

where *g* is the *acceleration of free fall.*

Potentiometer. An instrument, used to measure electric potential.

Pound. A British unit of *mass*. 1 Pound=0.45359237 *kilogram.*

Poundal. A unit of *force* in the *f.p.s. system.* It is the force which acting on a *mass* of I *pound*, will produce in it an *acceleration* of 1 foot per second.

Powder Metallurgy. The science of manufacturing small *metal* articles by sintering powdered metals under *heat* and *pressure.*

Power (math.). The number of times a quantity is successively multiplied by itself.

Power (phys.). The rate of doing *work*. The *SI unit* of power is the *watt.*

Power Alcohol. *Ethanol* used as a *fuel.*

Power Factor. In an electrical *circuit*, the ratio of the *power* disipated, *P*; to the product of the *electromotive force, E*, and the *current, I.*

Praseodymium. Pr. Element. R.a.m. 140.907. At No. 59, R.d. 6.78 m.p. 935°C., b.p. 3512°C. A rare soft silvery metal. Important ores arc *monazite* and *bastnasite.* Used in *Misch metal* and in some rare-earth *catalysts* for cracking oil.

Praxinoscope. Optical toy that gives effect of motion by reflection of successive pictures in a rotating box.

Precession of the Equinoxes. A slow movement to the west of the *equinoxes* about the *ecliptic*, caused by the *processional motion* of the *Earth.*

Precipitate (chem.). An *insoluble* substance formed in a *solution* due to a *chemical reaction.*

Precipitation (chem.). The formation of a *precipitate.*

Presbyopia. Long sight. A defect of vision normally occurring after the age of 45. The person is able to see distant objects clearly, but is unable to accommodate the eye to see near objects.

Pressure. The *force* per unit *area* acting on a surface. The *SI unit* of pressure is the *pascal* (Nm^{-2}). The *c.g.s. system* uses the *dyne* per square centimeter (1 Pa=10 dynes cm^{-2}). Other units are the *bar* (=10^5Pa), the *atmosphere* (=101.325Pa), the mm Hg(=133.322 Pa).

Pressurized Water Reactor PWR. A *nucler reactor* in which *water* is used as the coolant and the *moderator*, but in whict the water is maintained at a high *pressure* to prevent its *boiling.*

Primary Cell. Voltaic cell, galvanic cell. A device for producing an *electric current* as the result of a *chemical reaction.*

Primary Coil. Primary winding. The input coil of a *transformer* or *induction coil.*

Primary Colours (Phys.). Red, green, and a bluish-violet. Any *colour* may be produced by suitably combining *light* producing these. Also the *pigment colours* red, yellow, and blue, these cannot be imitated by mixing any other pigment colours.

Prime Number. A number which is not divisible by any whole number other than itself and one.

Primordial. Original ; existing from beginning.

Principal Plane (phys.). A *plane* containing the *optic axis* and either the *ordinary ray* or *the extraordinary ray.*

Principle Point (phys.). Two points on the *otpical axis* of a thick *lens* such that if the object distance is measured from one and the image distance from the other we get the *equations* for object-image distance, which are similar to those obtained for a thin lens.

Principal Section (phys.). A *plane* passing through the *optic axis* of a *crystal* that exhibits *double refraction* and is at right angles to one of the crystal surfaces.

Printed Circuit. An *electronic circuit* in which the wiring between components, and certain fixed components are printed on an insulating board.

Prism, Optical. A triangular *prism* made of *transparent* material *e.g., glass.*

Prismatic. In the shape of a *prism.*

Prismatic optical Instruments. Instruments (field-glasses, etc.) using a right-angled *prism* to invert the inverted image produced by the *objective.*

Probability, Mathematical . A mathematical expression of the chance that a specified event will occur.

Probability Distribution of Electrons. The *probability* that an *electron* within an *atom* will be at a certain point in *space* at a given time; it is measured by the magnitude of the square of the *wave function.*

Proboscis. Long external snout-like process in mouth parts of lepidopetrous insects.

Process Control. The control of complex chemical or industrial process using electronic means.

Producer Gas. A *fuel gas* consisting of *carbon monoxide* (25% - 30%) nitrogen (50%-55%), and hydrogen (10%-15%). *Hydrocarbons* and *carbon dioxide* will also be present.

Product (math.). The result of multiplication of two or more quantities together.

Progesterone ($C_{21}H_{30}O_2$). A white crystalline *steroid hormone*, m.p. 128.5°C. It prepares the reproductive organs of mammals for pregnancy and for protecting the embryo.

Program (Prgramme). The sequence of instruction fed into a *computer* so as to enable it to carry out a process.

Projectile. Refers to a body that is thrown or project.

Prolactin. Lactogenic hormone produced by anterior pituitary gland; it stimulates mammary glands for milk production in mammals.

Proline. A white crystalline amino acid found in most proteins, m.p. 220°C.

Promethium Pm. A *radioactive element* of the *lanthanide* series. At. No. 61. M.p. 1080°C.,b.p. 2460°C. Occurs as a fission product of uranium in *nuclear reactors*. The most stable *isotope*, promethium-145, has a *half-life* of about eighteen years.

Promoter. Substance which itself is weak catalyst but increase greatly catalytic activity of other catalyst.

Prompt Critical. Can sustain a *nuclear fission chain reaction* on the *prompt neutrons* alone; without contribution from *delayed neutrons*.

Prompt Neutrons. *Neutrons* resulting from *nuclear fission* that are emitted without measurable delay, *i.e.*, in less than a milionth of a second.

Proof Spirit. *Ethanol* containing 42.28% alcohol by *weight*, or 57.10% by *volume*, and having a *relative density* of 0.91976 at 60°F also defined as the weakest solution of alcohol that would fire *gunpowder* when brought into contact with it and ignited.

Proof Spirit, Degrees. The number of degrees under proof is the *volume* percentage of *water* in a *solution* regarded as containing *proof spirit* and water; degrees over proof is the volume increase obtained when 100 volumes of the spirit are dilluted with sufficient water to obtain proof spirit. Spirits are usually sold on the basis, '30° under proof' or '70° proof' both of which mean the same. Such spirit contains 57.1 × 70/100 =39.97% alcohol by volume.

Propanal. Propionaldehyde, propyl aldehyde. CH_3CH_2CHO. A colourless *aldehyde*, b.p. 48.8°C., used in the manufacture of *plastics*.

Propane (C_3H_8). A flammable *gas*. B.d.-42.17°C used as a *fuel* in the form of bottle gas.

Propanedioic Acid. Malonic acid. $CH_2(COOH)_2$. A white soluble *dibasic acid*, m.p. 135.6°C., used in the manufacture of *barbiturates.*

Propanoic Acid. Propionic acid. CH_3CH_2COOH. B.p. 141°C,. used as a bread additive as CH_3COONa.

Propanol. Propyl alcohol. Either of two isomers *n*-propyl alcohol, $CH_3CH_2CH_2OH$, a colourless *liquid*, b.p. 97.2°C., used as a solvent. Or isopropyl alcohol, $CH_3CHOHCH_3$, a colourless *liquid*, b.p. 82.4°C., used for manufacture of *propanone* (acetone), as a *solvent.*

Propanone. Acetone, dimethyl ketone, CH_3COCH_3. A colourless flammable *liquid* having a pleasant odour, b.p. 56.5°C., used as a *solvent* in making *plastics.*

Propellant. Explosive substance used for filling cartridges, shell cases, and *solid fuel rockets.*

Propenal. Acrolin, acrylaldehyde. $CH_2 : CHCHO$. A colourless flammable *liquid* having an irritating smell, b.p. 52.5°C., used to make *acrylic resins* and pharmaceutical products.

Propene. Propylene. $CH_2CH.CH_3$. A colourless *gas*, b.p.— 47°C.

Propenoic Acid. Acrylic acid. $CH_2:CHCOOH$. A corrosive liquid. m.p. 13°C., b.p. 141°C., used in the manufacture of acryeic resins.

Propenonitrile. Acrylonitrile, vinyl cyanide. $CH_2:CHCN$. A colourless flammable *liquid*, b.p. 78°C., used to prepare *acrylic resins*, synthetic rubbers, and synthetic fibres.

Proper Fraction. A fraction having *numerator* less than the *denominator.*

Proper Motion of the Star. Refers to the component of a *star's* motion in *space* relative to the *Sun* that is perpendicular to the line of sight.

Proportion (math.). An equality between two *ratios.*

Proportional Counter. A *counter tube* having the output pulse proportional to the number of *ions* produced.

Propyl. The *univalent radical* C_3H_7 —.

Propylene. *Propene.*

Prosthetic Group. A *non-protein* group that is combined to a protein.

Protactinium (Pa). *Radioactive* element. At. No. 91. The most abundant *isotope* has a *mass number* of 231 and a *half-life* of 32480 years.

Protargol. A powder consisting of finely divided silver and *proteins;* with *water*, it forms a *colloidal solution* of silver.

Proteases. Proteinases. A group of *enzyme* that are capable of breaking up *proteins* into *amino acids*, of building up amino acids into proteins, and of substituting one amino acid for another in protein *molecules* present in all living tissues.

Proteins. A class of complex nitrogenous *organic compounds* of high *relative molecular mass* (18 000–10 000 000). Protein *molecules* consists of hundreds of thousands of *amino acids* joined together by the *peptide* linkage into one or more interlinked *polypeptide* chains, folded in a variety of different ways. Most proteins form *colloidal solutions* in water or dilute *salt* solutions, but some (notably the fibrous proteins) are *insoluble. Enzymes* are a important group of proteins, they determine the *chemical reactions* likely to take place in a cell, and therefore the characteristics that it will have.

Proteolytic. (Proteoclastic). Capable of decomposing or *hydrolysing proteins.*

Protium. The hydrogen *isotope* having *mass number* of one.

Protolysis. A reaction involving the transfer of *protons.*

Proton. A stable *particle*, with *electric charge* equal in magnitude to that of the *electron* but of opposite sign and with *mass* 1836.12 times greater than that of the electron ($1.672\,614 \times 10^{-27}$ *kilogram*).

Proton Number. *Atomic number.*

Protoplasm. Living substance.

Provitamin. Substance capable of conversion into vitamin in course of digestion, carotene is provitamin of vitamin A.

Prussian Blue. Potassium iron (III) hexacyanoferrate (II). $KFe[Fe(CN)_6]$. A deep blue substance obtained by the action of a ferric *salt* on *potassium hexacyanoferrate* (II) (potassium ferrocyanide).

Prussic Acid. An intensely poisonous solution of *hydrocyanic acid*, HCN.

Psammon. Organisms that live between sand particles on lake or on sea shore.

Psammophil (e). Refers to sand loving plant.

Psammosere. Communities developing in sandy habitat during succession and ultimately stabilise such a habitat.

Pseudo-Acid. *Compound* that is not acid but transformable into acid.

Pseudoaromatic. A ring compound containing *conjugated double bonds* in the manner of an *aromatic* compound, having properties different to those of an aromatic compound.

Pseudoscalar. A *scalar quantity* that changes sign during the transition from a right-handed to left-handed system of coordinates.

Pseudovector. Axial vector. A *vector* quantity that changes sign during the transition from a right-handed to a left-handed system of coordinates.

Psi Particle (J. Particle). A *meson* having no charge but an anomalously long lifetime.

Psychrometry. Measurement of air humidity.

Psychrophilic. Micro-organism, it thrives below 20°C.

Pteroylglutamic Acid. Folic acid, medicinally active in pernicious anaemia.

Ptilosis. Plumage or mode of feathering.

Ptomaines. A class of extremely poisonous *organic compounds* formed during the putrefaction of *proteins* of animal origin.

Ptyalin. Enzyme present in saliva, it converts starch into sugar.

Ptyalism. Excessive flow of saliva.

Puddling Process. The preparation of nearly pure *wronght iron* from *cast iron* containing a high percentage of carbon. The cast iron is heated with *haematite*, Fe_2O_3, in this process.

Pulley. Wheel that turns on an axis and carries belt on its rim; used for lifting weights or changing direction of pull or transmission of power etc.

Pully and Belt. Mechanical arrangement to transmit torque from one shaft to another.

Pulmonary. Relating to lungs.

Pulmonary Artery. Paired artery that carries deoxygenated blood from right ventricle of heart to lungs.

Pulsars. *Stars* that emit *radio frequency electromagnetic radiation* at extremely regular intervals. It has been suggested that pulsars are *neutron stars*, emitting pulses of radiation when they rotate.

Pulse. A brief increase in the magnitude of a quantity whose value is generally constant (*e.g., current* or *voltage*).

Pulse Height Analyser. An instrument in an *electronic* circuit that permits only *voltage pulses* of predetermined *amplitudes* to be passed to succeding circuits.

Pulse-Jet. A type of *ram*-jet wherein the *combustion* process is not continuous, but occur at intervals between which the *pressure* in the combustion chamber is allowed to build up.

Pure Line. Formation of homozygous individuals which are identical to each other and continue to breed true.

Purine ($C_5H_4N_4$). White crystalline substance, forms uric and ($C_5H_4N_4O_3$) with oxygen.

Purple of Cassius. A purple *pigment*, made up of a *mixture* of colloidal gold and tin (IV) acid used for making ruby *glass*.

Push-Pull. An *electronic circuit* having two components that are out of phase by 180°.

Putrefaction. Type of bacterial decomposition of protein-containing substrate with formation of bad-smelling amines rather than ammonia.

PVC (Polyvinyl Chloride). Synthetic thermoplastic material.

Pyknometer (Pycnometer). Apparatus used for determining density and coefficient of expansion of liquid.

Pyorrhoea. Infection of gums and tooth sockets. It is followed by discharge of pus and loosening of teeth.

Pyrene. Hydrocarbon ($C_{16}H_{10}$), obtained by dry distillation of coal. Fruit-stone.

Pyrex. A type of *borosilicate glass.* It is resistant to heat and chemical attacks; widely used in laboratory glassware.

Pyridine (C_5H_5N). A colourless *heterocyclic liquid* having unpleasant smell. b.p. 115°C occurs in *bone-oil* and *coal-tar* used for making *methylated spirit.*

Pyridoxine. Vitamin B_6. $C_8H_{11}NO_3$. A *pyridine derivative*, a member of the *vitamin* B complex; it is a importance in the utilization of *unsaturated fatty acids* by many *organisms.*

Pyrimidine ($C_4H_4N_2$). An *organic base,* m.p. 22°C., b.p. 123.5°C., made up of a *heterocyclic* six-membered ring. *Derivatives* are of great biological importance, they occur in *nucleic acids e.g., Uracil, thymine* and *cytosine* are typical of such derivatives.

Pyrites. Natural *sulphides* of certain *metals.* Iron pyrites is FeS_2; Ccopper pyrites (fools' gold) is $CuFeS_2$.

Pyro. Prefix denoting fire, strong heat. In chemical nomenclature it denotes a substance obtained by heating; *e.g.*, pyroboric acid, obtained by heating *boric acid.*

Pyrocatechol. *1,2-dihydroxybenzene.*

Pyro-Electric. Becoming positively and negatively charged at opposite poles on heating and cooling.

Pyroelectricity. The peoperty of certain *crystals, e.g., tourmaline*, of acquiring *electric charges* on opposite faces when heated.

Pyrogallol. Pyrogallic acid, benzene -1,2,3-triol. $C_6H_3(OH)_3$. A white crystalline *solid*, m.p. 132°C., a powerful *reducing agent; alkaline solution* rapidly absorbs oxygen, used in photographic *developing* and in gas analysis for the estimation of oxygen.

Pyroligneous Acid. A watery *liquid* obtained by the *destructive distillation* of wood. It is a mixture of *ethanoic* (acetic) *acid*, CH_3COOH, *methanol*, CH_3OH, *propanone* (acetone), $(CH_3)_2CO$.

Pyrolusite. Natural *manganese (IV) oxide* (manganese dioxide; MnO_2). A black crystalline *solid*, r.d. 48; it is the *ore* of manganese.

Pyrolysis. Chemical *decomposition* by *heating*.

Pyrometers. Instruments used for measuring high *temperatures*.

Pyrometry. Measurement of high temperatures using pyrometer.

Pyrophoric Alloys. Alloys capable of emitting sparks when scraped or struck; flints in lighters.

Pyrophorus. Substance that takes fire on exposure.

Pyroscope. Instrument used for measuring intensity of radiant heat.

Pyrotechnics. Fireworks.

Pyroxenes. A group of *minerals* consisting of *silicates* of magnesium, iron, and calcium.

Pyrrole (C_4H_5N). A colourless liquid *heterocyclic* compound, b.p. 103°C., present in coal-tar.

Pyruvic Acid. 2-oxopropanoic acid. $CH_3.CO.COOH$. A *liquid organic acid*, m.p. 13°C. Pyruvic acid is broken down in the *citric-acid cycle*.

Pythagoras, Theorem of. In a right-angled *triangle* the square on the *hypotenuse* is equal to the sum of the squares on the other two sides.

Python. Very large non-poisonous snake that crushes its prey.

Pyuria. Refers to presence of pus in urine.

Pyxidium (pl. Pyxidia). Capsule that dehisces by transverse circular split, so that upper part of pericarp forms a lid.

Q

Q-band. Radio frequency band of 36 to 46 gigahertz.

Q-enzyme. Branching enzyme that is responsible for formation of bonds during synthesis of amylopectin.

QSG. Quasi stellar galaxy. A *quasar* that is not a radio source.

Quadrant. Quarter-circle. A *sector* of a *circle* bounded by an arc and two radii at right angles.

Quadrat. Square piece of vegtation, say one square metre, chosen at random to study composition of vegetation in that region.

Quadrate. Bone of suspensorium found in amphibia, birds and reptiles; helps to suspend lower jaw in authostylic jaw suspension.

Quadratic Equation. An *equation* involving the *square* or second *power* of the unknown quantity. Any quadratic equation may be written in the form

$$ax^2 + bx + c = 0.$$

Quadrature. Position of moon or an outer planet so that a line between it and earth forms a right angle with the line that joins earth to thesun.

Quadrilateral. Figure bound by four straight lines.

Quadrivalent. Showing a valency of four.

Quadruple Point. Unique condition of temperature and pressure where four phases of a two-component system exist in equilibrium.

Qualitative Analysis. Determination of chemical nature of substance.

Qualitative Characters. Characters of community as can be studied by subjective assessment.

Quantitative Analysis. Determination of quantities of substances present, by chemical methods.

Quantity of Electricity. The amount of *electric charge* flowing through a circuit; *i.e.*, the product of the current and the time for which it flows. The *SI unit is the coulomb.*

Quantized. Quantity that can only have certain discrete values (each of which is called a *quantum*). Such a quantity cannot vary continuously, differences in value being separated by 'jumps'.

Quantum. *Energy* exists in discrete units, only whole numbers of which can exist; each unit is called a quantum (plural 'quanta'). The quantum of *electromagnetic radiation* is the *photon.*

Quantum Numbers. *Integral* or half-integral numbers that specify the state of a system or its components in *quantum mechanics.* An *electron* within an atom, for example, is specified by four quantum numbers in the *Bohr theory.*

Quantum Theory. The theory based on the concept of the discontinuity of *energy.* The system of *quantum mechanics* evolved from this theory.

Quark. Hypothetical fundamental particle(s) that are postulated to be building blocks of elementary particles.

Quart. Unit of capacity. 1 Quart = $\frac{1}{4}$ gallon.

Quartz. Natural crystalline form of silica (SiO_2).

Quartz Clock or Watch. A clock or watch regulated by a quartz crystal. Being much more accurate than a balance-wheel or pendulum regulated clock, it is now used in all accurate clocks and watches.

Quartz Lamp. Mercury vapour lamps with transparant envelope made of quartz; capable of passing out even ultra-violet rays that are absorbed by ordinary glass.

Quasars. Quasi stellar radio sources. They were originally located by radio astronomers because of their powerful *radio frequency* emissions.

Quaternary Ammonium Compounds. *Compounds* having general formula NR_4OH ; they are derived from *ammonium hydroxide* NH_4OH, by replacement of the hydrogen *atoms* by *organic radicals.*

Quenching. Process of terminating the discharge in Geiger counter by preventing re-ignition.

Quenching of Steel. Rapid cooling by immersing into water or oil to harden the steels.

Quicklime. Calcium oxide (CaO).

Quiescent. Refers to inactive, resting, non-feeding stage of an organism.

Quiet Sun. The *Sun's* condition when no *sunspots, solar flares*, or *solar prominences* are taking place. *Radio-frequency* emission from the Sun, which has to be observed during the rare periods of the quite sun. It has enabled measurement of temperature of various layers of the solar atmosphere.

Quinhydron [$C_6H_2(OH)_2.C_6H_4O_2$]. Green crystalline substance used in photography and as antioxidant.

Quinic Acid. Acid obtained from cinchona bark.

Quinidine ($C_{20}H_{24}N_2O_2$). Colourless crystaline substance isomeric with quinine; used in medicines.

Quinine ($C_{20}H_{24}O_2N_2\ 3H_2O$). Colourless, bitter tasting, crystalline alkaloid, obtained from cinchona bark; used as effective medicine against malaria.

Quinoline (C_9H_7N). Colourless liquid base, boiling point 237°C, present in coal-tar; used as solvent and in manufacture of dyes.

Quinones. Series of aromatic compounds in whose molecules two hydrogen atoms in same benzene nucleus are replaced by oxygen atoms, forming carbonyl groups.

Quinquevalent. Pentavalent: having valence of five.

Q-unit. Unit of energy used in measuring heat energy of fuels reserves ($= 1.055 \times 10^{21}$ joules).

Q-value. Amount of energy released in a nuclear reactor.

Q-wave. Surface wave having horizontal motion, its velocity depends on density and rigidity modulus.

R

Racemase. Group of enzymes which catalyse racemisation reactions.

Racemic Acid. Optically inactive form of tartaric acid; present in grapes.

Racemic Mixture Racemate. An *equimolecular mixture* of the two *optically active* forms of a substance. It is denoted by the *dl., e.g., dl-tartaric acid*; it is optically inactive and is externally compensated.

Racemisation. Process of conversion of optically active compound into optically inactive form by chemical methods.

Rad. Measure of the amount of energy absorbed from radiation by tissue or other biological material.

Radar. Radio angle detection and range instrument to locate objects (aeroplanes, missiles, satellites, ships etc.), measuring distance and latitude etc. It uses beamed and reflected radio frequency energy.

Radial Symmetry. Similar parts arranged around a common centre as in starfish, sponge etc.

Radian (Rad). SI unit of plane angle (=57.296°).

Radiance L. The *radiant intensity* of a source in a given direction, per unit transverse area.

Radiant Energy. Energy transmitted in the form of *radiation.*

Radiant Flux Φ_e. The total power emitted or received by a body in the form of *radiation*, measured in *watts.*

Radiant Intensity I_e. The *radiant flux* per unit *solid angle* emitted by a point source.

Radiata. Lower animals with basic radical symmetry. They include, coelenterates; ctenophores and sponges.

Radiation. Transmission of energy as wave or in form of stream of moving particles like α-rays or β-particles etc.

Radiation Hazard. The likely danger to health due to exposure to *ionizing radiation* or the consumption of a *radioactive* substances.

Radiation Potential. Resonance potential. The *energy* required to transfer an *electron* from its normal position in an *atom* to some other position; *i.e.*, to an *energy level* of higher energy.

Radiation Pressure. The pressure exerted on a surface by *radiation.* The pressure of the Sun's radiation on the Earth is about 4×10^{-6} pascal.

Radiation Sickness. Illness due to exposure to *ionizing radiation.* It causes vomitting and diarrhoea, followed in some cases by leukaemia.

Radiation Temperature. The surface temperature of a *star* or other celestial body, it is sometimes called the optical temperature.

Radiative Collision. A collision between charged particles in which part of the *K.E.*, is converted to electromagnetic radiation.

Radiator. Small tank-like part attached to engines of buses, cars, trucks etc., it serves to keep engine cool. When filled with water.

Radical (Radicle). Group of atoms that behave like single atom. Part of embryo which develops into root. Leaves which arise from back of plant at ground level.

Radio. Equipment used for receiving communication or programmes transmitted from control stations through electromagnetic waves; wireless transmission. It also includes telegraphy, telephony, television, radar etc.

Radioactive. Relating to *radioactivity,* radiocarbon is an abbreviation for radioactive carbon.

Radioactive Age. The age of a *mineral fossil,* or any wooden object as estimated from its content of *radionuclides.*

Radioactive Equilibrium. A state finally reached when a *radioactive* substance of slow *decay* yields a radioactive product on *disintegration.*

Radioactive Series. A series of *radionuclides,* in which each except the first being the *decay* product of the previous one. The three naturally occurring series are the thorium series (starting with thorium-232), the actinium series (uranium-235), and the uranium series (uranium-238). The final member of each series is an isotope of lead.

Radioactive Standard. A specimen of a material containing a *radionuclide* of precisely known rate of *decay* which is used for the calibration of instruments measuring *radiation.*

Radioactive Tracing. A method for tracing the course of an element through a biological, chemical, or mechanical system by introducing a small amount of a *radioisotope*, known as tracer, the path taken by the stable isotope of the same element can be followed.

Radioactivity. The process of spontaneous disintegration of an unstable atomic nucleus accompanied by emission of alpha, beta and gamma radiations.

Radio Astronomy. Study of celestial bodies by reception and analysis of electromagnetic radiation which are reflected by them, astronomical study using radar.

Radio-beacon. Apparatus that transmits signals which are for direction finding.

Radiobiology. Branch of biology dealing with effects of radiations on living organisms.

Radiochemistry. Chemistry of radioactivity.

Radio-compass. Instrument for finding wireless direction.

Radio-diagnosis. Branch of medical radiology which deals with the application of X-rays to diagnosis.

Radio-element. Radioactive isotope of an element.

Radio Frequency. The *frequency* of *electromagnetic radiation* within the range 10 kilohertz to 100 000 megahertz used in Radio.

Radio-frequency heating. Industrial *induction* or *dielectric heating,* specially when the *frequency* of the alternating field is above 25 *kilohertz.*

Radiogenetics. Scientific study of various effects of radiation upon heredity.

Radiography. X-ray photography.

Radio-isotope. Radioactive isotope of any stable element.

Radiolaria. Order of marine protozoa having final radical pseudopodia.

Radiology. Study of radioactivity and radiation alongwith their application to medicine.

Radiolucent. Almost *transparent* to *radiation,* especially *X-rays* and *gamma rays.*

Radioluminescence. *Fluorescence* arising from *radioactive decay.*

Radiolysis. The chemical *decomposition* of substances due to *radiation.* The radiation causes *ionization* and *excitation* which promote further reactions.

Radiomicrometer. A very sensitive instrument used for measuring *heat radiations.* It consists of a *thermocouple* connected directly into a single copper loop that form the coil of sensitive *galvanometer.*

Radionuclide. A *nuclide* of *radioactive atom.*

Radioopaque. Not permitting radiation to pass through it.

Radiophone. Instrument that produces sound by radiant energy; wireless (cordless) telephone.

Radioscopy. Examination with the help of X-rays.

Radiosonde. A small balloon to carry meteorological instruments into the *Earth's atmosphere.*

Radio Source. A source of *electromagnetic radiation* of *radiofrequencies* outside the *solar system.* Techniques of *radio astronomy,* both within the *Galaxy* and outside it, but only a small number have been identified with *stars* that can be located with *optical telescopes.* Other sources are *supernovae* explosions and remnants, colliding *galaxies* and gas clouds, *quasars,* and *pulsars*; some sources, however, remain unexplained.

Radio Telegraphy. Wireless telegraph.

Radio Telephony. The use of *radio,* instead of wires for all or part of a *telephone system.*

Radio Telescope. Instrument used in *radio astronomy* to pick up and analyse the *radiofrequency electromagnetic radiations* of extraterrestrial sources.

Radiotherapy. Treatment of disease by radiation (X-rays etc.).

Radiowave. Electromagnetic wave that is produced by reversal of current in a conductor; used for broadcasting and television transmission.

Radium (Ra). White radioactive metallic element discovered by Curies (1898); b.p. 1140°C, oxidises violently with air and reacts with water liberating hydrogen that readily catches fire; used in manufacture of photoelectric cells, radio tubes, alloys etc.

Radius. Shorter and thicker of two bones of forearm in vertebrates located on thumb side. Shortest distance from centre to periphery of a circle.

Radius of Gyration. The *moment of inertia* I, of a body of *mass* m about a given axis can be expressed in the form $I=mk^2$, k is known as the radius of *Gyration* about the axis

Radius Vector (astr.). A line drawn from a central body to a *planet* in any position in its *orbit.*

Radius Vector (math.). The position of any point *P* in space with respect to a given *origin O* may be completely defined by the direction and length of the line *OP*. This line is called the radius vector of the point *P*.

Radix. A number that forms the *base* of a system of numbers, *logarithms*, etc., *e.g.*, the radix of the *binary notation* is 2.

Radon (Rn). Colourless; monoatomic, gaseous radioactive element last member of zero group, obtained by dissolving salt in water and allowing solution to stand for about a month; used in radiotherapy.

Raffinate. Refined liquid after its soluble components have been removed by solvent extraction.

Raffinose. ($C_{18}H_{32}O_{16}5H_2O$). Melitose; colourless crystalline trisaccharide, it occurs in beet-sugar but does not have sweet flavour.

Rain. Water from clouds falling in continuous drops; condensed moisture of atmosphere falling in drops.

Rainbow. Seven coloured arch formed in sky opposite sun by refraction and internal reflection of light in raindrops. It is generally seen as huge semi-circle.

Raingauge. Equipment used for recording rainfall at a particular place.

Raman Effect. Monochromatic light when passed through transparent medium is scattered into original wave-lengths as well as larger wave- lengths (Raman lines); useful in study of molecular energy level of liquid.

Ram Jet. A simple type of aerodynamic *reaction propulsion* system wherein *thrust* is obtained by the *combustion* of *fuel* in air, compressed only by the forward *speed* of the vehicle. A ram jet is also known as a 'flying drainpipe'. A ram jet has to be launched at high speed and is unable to take off unaided from rest.

Ramsden Eye-Piece. An *eye-piece* consisting of two *plano-convex lenses* (curved surfaces inwards) of equal *focal length* f, and separated by a distance of $2f/3$.

Ramus. Branch of any structure,—Process of Bone, feather barb, nerve etc.

Random Sample. A sample taken in such a way that every individual, object, or component to be sampled, has an equal *probability* of forming part of the sample.

Raney Nickel. Spongy nickel, used as catalyst in hydrogenation of fats and oils. May be obtained by dissolving Al in Ni—Al alloy with NaOH.

Rankine Temperature **°R.** The absolute *Fahrenheit scale.* Zero degrees Rankine is —459.67°F. and therefore °F + 459.67 =°R.

Raoult's Law. When a *solute* that does not dissociate in *solution* is dissolved in a *solvent* to form a *dilute* solution, then the decrease in *vapour pressure* to the original vapour pressure is equal x_2. Where x_2 is the mole fraction of solute.

Raphe. Vertical marking on seed coating. It is developed from antropous ovule.

Raphis. Calcium oxalate crystals. They are found to occur in bundles in some plants.

Rarefaction. A reduction in *pressure.*

Rare Gases. *Insert gases.*

Raster. The pattern of lines capable of scanning the fluorescent screen of a *cathode-ray tube* in a *television* receiver.

Ratio. The numerical relation one quantity bears to anther of the same kind, *e.g.* 60 tons and 40 tons, and 30 and 20, are both in the ration of 3 : 2.

Rationalized Units. Units of measurement, such as *SI units.*

Rational Number (math.). A whole number, or a number that can be expressed as the *ratio* of two whole numbers.

Ray. The *rectilinear* path along which any *radiation,* travels in any direction from a point source.

Rayon. Formerly 'artificial silk', the term is now restricted to two types of man-made *cellulose* fibres : *viscose* rayon.

React. Respond to stimulus influence etc.

Reactant. Substance taking part in chemical reaction.

Reaction Propulsion. Jet propulsion. A form of propulsion of which a high-speed stream of *gas* reacts upon the vehicle in which it was produced in accordance with *Newton's* (third) *law of motion,* so that the vehicle is propelled through the medium in which it is travelling. It is the only known method of propulsion through *space* where there is no supporting medium and it is upon this principle that *rockets* are propelled.

Reactive (chem.). Readily participating into *chemical reaction.*

Reactive Dyes. *Dyes* that react chemically with the substances to be dyed. They are used for cellulose fibres.

Reactor (chem.) Any vessel in which a *chemical rection* is carried out.

Reactor (phys.). A device for introducing reactance into an electrical circuit.

Reagent. A chemical substance that produces a *chemical reaction.*

Realgar. Natural red *tetrarsenic tetrasulphide* [arsenic (II) sulphide], As_4S_4.

Real-Time Working. A method of operating a *computer* in which information from the computer output is available at the time it is required by the rest of the system.

Reaumur Scale. A *temperature* scale in which the *melting point* of *ice* is taken as $0^{\circ}R$ and the *boiling point* of *water* as $80^{\circ}R$.

Rectangle. A *quadrilateral* having right angles between all four sides.

Rectification (chem). The purification of *liquid* by *distillation.*

Rectification (maths.). The process of finding the length of a curve.

Rectification (phys.). The conversion of an *alternating* into a *direct current.*

Rectified Spirit. *Ethanol*, usually an ethanol/water mixture containing 95.6% ethanol.

Rectifier (phys.). A device for transforming an *alternating current* into a *direct current.*

Rectifying Valve. The *thermionic valve* used for *rectification* is the *diode.*

Rectilinear. In a straight line; consisting of straight lines.

Rectilinear Propagation of Light. *Light* travels in straight lines.

Red Blood Corpuscles (RBCs). Blood cells having smooth-surface, flattened discs, round to oval in shape and contain haemoglobin. In human blood these are roughly five millions per cubic millimetre and carry nearly all oxygen contained in blood.

Red Lead (Pb_3O_4). Minium, bright scarlet powder; used as pigment in glass and as an oxidising agent.

Redox Chain. Oxidation-reduction system in which one substance is oxidised and another reduced.

Redox Reaction. Oxidation-reduction reaction. A *chemical reaction* in which oxidation and reduction occur simultaneously.

Red Shift. A *Doppler effect* in which the displacement of spectral lines takes place towards the red end of the *spectrum* observed in the *light* from certain *galaxies* is taken to mean that they are receding into *space*, as a consequence of the *expansion of the Universe.*

Reduced Equation. An *equation of state* of a *gas* in which the temperature, pressure, and volume are replaced by *reduced temperature, pressure* and *volume respectively.*

Reduced Temperature, Pressure and Volume. Ratios of the *temperature,* the *pressure,* and the *volume* to the *critical temperature, critical pressure,* and *critical volume* respectively.

Reducing Agent. A substance that removes oxygen from, or adds hydrogen to, another substance.

Reductase. An *enzyme* which promotes a *reduction* reaction.

Reduction. Removal of oxygen from a substance, or the addition of hydrogen to it.

Redundancy. A term used in *information theory.*

Re-Entry. The position, time, or act of re-entering the *Earth's* atmosphere after a journey into *space.*

Refine, Purify. Removal of impurities.

Reflectance. It is the *ratio* of the intensity of the reflected radiation to the intensity of the incident radiation.

Reflecting Telescope. Reflector.

Reflection, Angle of. The angle between a reflected ray from a surface, and the *normal* to the surface at that point.

Reflection of Light, Laws of. 1. The incident *ray,* the reflected ray, and the *normal* to the reflecting surface at the point of incidence lie in the same *plane.* **2.** The *angle* between the incident ray and the normal, (*i.e.*, the angle of *incidence*) is equal to the angle between the reflected ray and the normal, (*i.e.*, the angle of *reflection*).

Reflector. Surface, object or device that reflects radiant energy as light, sound etc.

Reflex. Involuntary (quick, automatic) response to stimulus.

Reflex Angle. An *angle* greater than 180° and less than 360°

Reflex Camera. A *camera* that allows the photographer to view and focus the exact scene he is photographing.

Reflex Condenser. A type of *condenser* in which the *vapour* over a *boiling liquid* is condensed to a liquid, that flows back into the vessel, thus preventing its contents from boiling dry.

Reforming. A catalytic reaction in which straight-chain *alkanes* are converted into branched-chain compounds or *aromatics.* It is used to produce *petrol* from crude oil.

Refraction, Angle of. The angle between the refracted ray and the *normal* to the surface at the point of *refraction.*

Refraction, Correction. The correction needed to the observed *altitude* of a heavenly body due to the *refraction* of the *light* it emits or reflects by the Earth's *ătmosphere.*

Refraction of Light. When a *ray of light* travels obliquely from one medium to another, it is bent or refracted at the surface separating the two media.

Refractive Index of a Medium *n*. The ratio of the sine of the angle of *incidence* to the sine of the angle of *refraction* when *light* is refracted from a *vacuum* into the medium.

Refractivity. If the *refractive index* of a medium is *n*, its refractivity is defined as *n*—1.

Refractometer. Apparatus used for measurement of refractive index of a substance.

Refractory. Resistant to ordinary treatment, stimulus, disease etc. Substance that can resist high temperature; used as bricks for lining furnaces.

Refrigerant. Cooling or freezing agent–liquid nitrogen etc.

Referigerating Cycle. The cycle of operations that takes place in a refrigerator.

Refrigerator. Apparatus used to produce and maintain low temperature.

Regelation of Ice. Ice near its melting point is melted by sufficient pressure, and solidification or regelation takes place again on removing pressure.

Regenerator. A *heat exchanger* consisting of a chamber filled with bricks arranged in checkerwork. *Heat* from the exhaust gases is stored in the brickwork and transferred to the cold air, so increasing the efficiency of the combustion process.

Reindeer. Large heavy deer, antlered in both sexes.

Relative Density (r.d.). Specific gravity; ratio of density of solid or liquid at specified temperature (usually 20°C) to density of water at temperature of its maximum density (4°C).

Relative Humidity (RH). Hygrometric state of atmosphere; percentage of water vapour in air at a given temperature as compared to total amount of water the air can hold at that temperature.

Relative Molecular Mass. Molecular weight. The ratio of the average *mass* per *molecule* of a substance to 1/12 the mass of a carbon-12 *atom.* It is equal to the sum of the *relative atomic masses* of the atoms of the molecule.

Relative Molecular Mass Determination. Relative molecular mass can be determined by measuring *vapour density, depression of freezing point, elevation of boiling point* and *osmotic pressure.*

Relativistic Mass. The *mass* of a body that is travelling at a speed comparable to the *speed of light.* The relativistic mass, *m*, of a body travelling at a speed, *v*, is given by:

$$m=m_o\left(1-\frac{v^2}{c^2}\right)^{-1/2}$$

where m_o is the rest mass and *c* is the speed of light.

Relativistic Mechanics. The *mechanics* that replaces *Newtonian mechanics* for bodies travelling at *relativistic speeds.*

Relativistic Particle. A particle having a speed comparable to the *speed of light.*

Relativistic Speed. A *speed* nearly equal to the *speed of light*, at which the effect of the theory of *relativity* is significant.

Relativity Theory. A theory, put forward by Einstein, it recognizes the impossibility of determining absolute motion and leads to the concept of a four-dimensional *space-time continum.*

Relay, Electrical. A device by which the *electric current* flowing in one circuit can open or close a second circuit and thus control the switching on and off of a current in the second circuit.

Reluctance. Magnetic resistance. The ratio of the *magnetomotive force* acting in a *magnetic circuit* to the *magnetic flux*, unit is henries^{-1}

Reluctivity. *Reciprocal* of *magnetic permeability.*

Rem. The unit dose of *ionizing radiation* that produces the same biological effect as that due to one roentgen of *X-rays.*

Remanence. Retentivity. The residual *magnetic flux density* of a *ferromagnetic substance* subjected to a *hysteresis cycle* when the magnetizing field is reduced to zero.

Renin. Hormone produced by kidney causes increases in blood pressure by its role in production of angiotensin.

Rennet. An extract of the fourth stomach of the calf. It contains rennin.

Rennin. Enzyme produced in stomach of mammals, it coagulates proteins in milk.

Repellent. Distasteful, disagreeable.

Resins. *Amorphous organic compounds* secreted by certain plants and insects; they are usually *insoluble* in *water* but *soluble* in various *organic solvents*. Typing natural resins are *rosin* and *shellac. Synthetic* resins are a group of synthetic substances whose properties resembled natural resins.

Resistance. Capacity to withstand toxic effects of pesticide.

Resistance Thermometer. Accurate *thermometer*, in which the temperature is deduced from the measurement of the resistance of a spiral of a metal (usually platinum) in the form of a wire.

Resistivity. Specific resistance. It is equal to the *reciprocal* of its *conductivity*. The resistivity, ρ equals RA/l where R is the resistance of a uniform *conductor* of length l and cross-sectional area A. It is usually expressed in *ohm metres.*

Resistor. A device used in *electronic* circuits primarily for its *resistance.*

Resolution of Vectors. The division of *vectors* into components acting in specified directions, perpendicular to each other.

Resolving Power. The ability of an optical system to produce separate images of objects very close to each other.

Resonance (chem). The resonance method of describing approximately the actual structure of a compound use a number of classical structures ('canonical forms') in terms of which the actual structure (the 'resonance hybrid') is described.

Resonance Nuclear. Resonance is said to occur in *nuclear reactions* if the *energy* of an incident particle or *photon* is equal, to the value of an appropriate *energy level* of the compound *nucleus.*

Resonant Cavity. A space enclosed by electrically conducting surfaces, in which electromagnetic *energy* may be stored or exited.

Resonant Circuit. An electronic circuit containing *resistance, inductance*, and *capacitance.*

Resorcinol Benzene-1, 3-diol, $C_6H_4(OH)_2$. A solid *dihydric phenol* m.p. 110°C., used in tanning and as an *intermediate* in the manufacture of *resins drugs*, etc.

Respiration. Respiration is the process by which living *organisms,* or their components, take oxygen from the *atmosphere* to oxidize their food and obtain *energy.*

Respiratory Pigment. A substance formed in *blood cells* or *blood plasma* which is capable of combining loosely and reversibly with oxygen, *e.g., haemoglobin.*

Respiratory Quotient (RQ). The Ratio of the *volume* of carbon *dioxide expired* by an *organism* or tissue to the volume of oxygen consumed by it over the same period.

Response. Activity due to change in internal or external environment.

Rest Energy. The equivalent of the *rest mass* of a body expressed in *energy* units, *i.e.*, m_0c^2, where m_0 is the rest mass and *c* is the *speed of light.*

Restitution, Coefficient of e. A measure of the *elasticity* of bodies upon impact.

Rest Mass. The *mass* of a body when at rest relative to the observer.

Resultant (Phys). A single *force* or *velocity* which produces the same effect as the two or more forces or velocities acting simultaneously.

Retardation (Phys.). Deceleration. The rate of decrease of *velocity* or *speed.*

Retina. Innermost layer of eye containing receptor cells.

Retort (Chem.). A *glass* vessel consisting of a large bulb with a long neck narrowing somewhat towards the end.

Retro-Rocket A small *rocket* that froms part of a larger one, it produces *thrust* in the opposite direction to that of the main rocket so as to decelerating it; *e.g.*, to enable a lunar *module* to make a 'soft' landing on the *Moon.*

Reverberation Time. The time taken for a *sound* to fall by 60 *decibels* in an auditorium.

Reverberatory Furnace. A furnace for operations in which mixing of material with full is not desirable the roof is heated by *flames*, and the *heat* is radiated down to the material off the roof.

Reverse Osmois. A process of *desalination* in which brine and pure *water* are separated by a *semipermeable membrane.*

Reversible Process (in *thermodynamics*). A reversible process can take place only in infinitesimal steps about equilibrium states of the system.

Reversible Reaction. A *chemical reaction* that may be made, to proceed in either direction under suitable conditions.

Reynolds Number. (*Re*). A dimensionless quantity applied to a fluid flow. For a *liquid* flowing through a tube,

$$(\mathrm{Re}) = u\rho l/\eta,$$

where u = *speed* of flow, ρ=*density* of the liquid, l=the diameter of the tube, and η = the coefficient of *viscosity* of the liquid.

Rhe. The unit of *fluidity raciprocal* of the *poise.*

Rhenium (Re). Element R.a.m. 186.20. At. No. 75. A hard grey *metal,* r.d. 20.53, m.p. 3180°C., b.p. 5627°C., used in *thermocouples* and as *catalyst.*

Rheobase. Minimum strength of electric current required to produce stimulation.

Rheology. The study relating to the deformation and flow of *matter.*

Rheopexy. The acceleration of a thixotropic increase of *viscosity* by gentle stirring.

Rheostan. An *alloy* of 52% copper, 25% nickel, 18% zinc, and 5% iron, used for electrical resistance wire.

Reostate. A variable electrical *resistor.*

Rhesus. Indian monkey.

Rhesus Factor. Rh. factor. A group of *antigens* in the red *blood cells* of some humans (said to be Rh. positive) but absent in some individuals (Rh. negative) (Rh. negative).

Rh. Factor. Group of antigens usually present in human blood which may cause dangerous reactions especially during pregnancy. Persons having this factor are Rh-positive, those lacking are Rh-negative. In India 97% people of are Rh-positive and only about 3% are Rh- negative.

Rhinoscope. Instrument used for examining nose.

Rhodium (Rh). Silvery - white metal that occurs alloyed with platinum; boiling point 3720°C; used in making scientific instruments.

Rhodophyta. Red algae ; highest group of algae characterised by red colour; contains pigments phycoerythrin and phycocyanin.

Rhodopsin. Pigments in rods of retina, essential for twilight vision.

Rhombus. A *quadrilateral* having equal sides.

Rh Scale. A scale of hydrogen pressures that gives a measure of the strength of a *reducing agent.*

Riboflabin. Water soluble vitamin of B-group (B_2); forms part of coenzyme concerned in cellular oxidation.

Ribonuclease. An *enzyme* which *catalyzes* the *hydrolysis* of *ribonucleic acid.*

Ribonucleic Acid (RNA). Long thread like *molecules* made up of single *polynucleotide* chains.

Ribose ($C_5H_{10}O_5$). A *pentose,* m.p. 95°C; the *dextrorotatory* form of great biological importance because it occurs in the *nucleotides* of *ribonucleic acid.*

Ribosomes. Small granules (about 10^{-8} *metre* in diameter) which occur in the *cytoplasm* of *cells* and are the sites of *protein* synthesis.

Richardson Equation. Richardson-Dushman equation. The equation that relates the number of *electrons* emitted by a heated metal surface to the *thermodynamic temprature, T,* in *thermionic emission.*

Ricinoleic Acid ($C_{17}H_{32}OHCOOH$). Yellow liquid which occurs in castor oil and boils at 227°C; used in manufacture of soaps.

Ring Compound. Chemical compound in whose molecule some or all atoms are linked in closed ring.

Ringer's Fluid. *Physiological* saline containing sodium, potassium, and calcium *chlorides*; widely used for sustaining *cells* or tissues during *in vitro* biochemical experiments.

RMS Value. *Root-maen-square value of a variable.*

RNA (Ribose Nucleic Acid). One of the nucleic acids present in single strand, it has pentose sugar ribose, phosphoric acid and four heterocyclic bases.

Roast. Heating ore to effect some chemical change which will facilitate smelting.

Rochelle Salt ($KNaC_4H_4O_6$). Potassium sodium tartrate; white crystalline soluble salt having piezoelectric properties; used in baking powder.

Rochon Prism. *Device* used for obtaining plane-polarized light. It consists of two prisms made of *quartz,* one cut parallel to the *optic axis* and the other perpendicular to it. Used for work with *ultraviolet radiation.*

Rock. In the scientific sense, a rock is any distinct material present in the Earth's crust but, in distinction from a *mineral,* it need not have a definite chemical composition and may consist of more than one mineral. A rock need not necessarily be hard or stone-like, *e.g.*, *clays* are regarded as rock materials. Rocks are classified as *igneous rocks, metamorphic rocks,* or *sedimentary rocks.*

Rock Crystal. A pure natural crystalline form of *silica,* SiO_2.

Rocket. A projectile driven by *reaction propulsion* that contains its own *propellants.*

Rocks Salt, Halite. Natural crystalline *sodium chloride,* NaCl.

Rod. A photosensitive cell in the retina of vertebrate eyes. Rods contain the pigment *rhodopsin.*

Rodenticide. Chemical used for killing rodents, rats and mice.

Rodinal. Photographic developer. An alkaline solution of 4-aminophenol, ($NH_2C_6H_4OH$) and sodium bisulphite ($NaHSO_3$).

Roentgen Rays. Name given by German physicist Konard von Roetgen to X- rays.

Roentgen (R). Unit of ionising radiation.

Rongalite. A *compound* of sodium sulphoxylate and *methanal,* $NaHSO_2$. HCHO. Used as a *reducing agent* in dyeing.

Root (Math). **1.** One of the equal *factors* of a number or quantity. The root of an *equation* is value of the unknown quantity that satisfies the equation.

Root-mean-square Value of Variable RMS. The square root of the average of the squares of a number of values, given by:

RMS = √ (Sum of squares of the individual values of the variable) /(Total number of values).

Rose's Metal. An *alloy* of 50% bismuth, 25% lead, and 25% tin; m.p. 94°C.

Rosin Colophony. A yellowish *amorphous resin* left as a residue from the *distillation* of *turpentine*. R.d. 1.08, m.p. 120°— 150°C. Used in varnishes, *soap*, and *soldering* fluxes.

Rotameter. An instrument for measuring the rate of flow of *fluids.*

Rotary Converter. An *alternating-current electric motor* mechanically coupled to a *direct-current generator*. Used for converting an AC supply into DC.

Rotascope. Instrument used to observe machinery in rapid movement.

Rotation. The angular displacement of a body.

Rotor. Rotating part of electric motor, dynamo or generator; armature.

Rubber. Tough elastic substance, made from milky liquid (latex) obtained by tapping certain trees; used to make products that require elasticity and moldability—tyres, tennis balls etc.

Rubber, Synthetic. Synthetic *elastomers* made from *polymers* or copolymers.

Rubidium (Rb). Element. R. a.m. 85.47. At. No. 37. A soft *reactive* white *metal* resembling soldium. R.d. 1.53, m.p. 38.9°C., b.p. 688°C. Occurs in a few rare *minerals.*

Rubidium-strontium Dating. A method of *dating* some rocks, used for specimens over 10^9 years old. It is based on the *decay* of the *radioisotope* rubidium-87 (*half-life* 5×10^{10} years) to yield strontium-87.

Ruby. A red form of *corundum*, Al_2O_3, its colour is due to traces of chromium. Used in *lasers* and as a gem stone.

Rudimentary. Body part that is not fully developed; reduced and functionless part of body.

Rugose. Wrinkled surface; covered with sunken lines.

Rules of Fajans. Fajans' rules state that an electrovalent bond will be replaced by a covalent bond if. (1) the charge on either of the *ions* resulting from an electrovalent donation of *electrons* is large (*i.e.*, if more than 1 or 2 electrons are donated); or (2) the volume of the *cation* is small or that of the anion is large.

Rusma. Depilatory of lime and orpiment.

Rust. Hydrated iron oxide ($Fe_2O_3.H_2O$), formed on surface of iron on being exposed to moisture and air.

Rut. To be in heat; periodic sexual excitement in male animals particularly stag.

Ruthenium (Ru). Metallic element belonging to platinum group, hard and brittle and not easily fused, boiling point 3900°C; strong catalyst.

Rutherford (rd). Unit of radioactive disintegration (rd=million disintegrations a second).

Rutile. Crystalline natural titanium dioxide (TiO_2).

Rutin. Drug used against fragility of small blood vessels.

Rydberg Constant. A constant relating to those atomic *spectra* that are similar to the hydrogen *atom* spectrum (see *Balmer series*). The Rydberg constant for hydrogen is 1.09677×10^7 m^{-1}. The general Rydberg formula is :

$1/\lambda = R\,(1/n^2 - 1/m^2)$,

where R is the Rydberg constant and n and m are positive *integers*.

S

Saccharase. Enzyme which catalyses hydrolysis of disaccharide to monosaccharides — sucrose to dextrose and levulose.

Saccharide. Simple sugar, monosaccharide.

Saccharimeter. Polarimeter, instrument used for testing sugars.

Saccharine (e) ($C_7H_4SO_3NH$). Extremely sweet white crystalline solid obtained from toluene but without calorific value, used as substitute of sugar.

Saccharometer. Hydrometer used for measuring concentration of sugar solutions.

Sacrificial Protection. The protection of ferrous metals against *rusting* using more reactive metal. For example, if *galvanized iron* is scratched so that the iron is exposed, it does not rust as the zinc *ions* go into solution before the iron ions.

Sacrum. It is the region in vertebral column that lies between lumber and column that lies between lumbe caudal regions.

Safety Glass. Glass so treated that it does not break or splinter on impact. Toughened glass is heat treated. Laminated glass has a thin layer of transport plastic sheet between two sheets of glass. Wire glass has a wire mesh built into it.

Safety Lamp (Davy Lamp). An oil-lamp that will not ignite flammable *gases, e.g., methane* (fire-damp).

Safrole. $CH_2.CHCH_2C_6H_3O_2CH_2$. A yellowish crystalline substance m.p. 11.2°C., b.p. 234.5°C., used in the manufacture of perfumes, flavours and *soaps.*

Sal Ammoniac. Ammonium chloride (NH_4Cl).

Salicin. $CH_2OHC_6H_4OC_6H_{11}O_5$. A colourless *soluble glucoside.* m.p. 200°C., used as an *antipyretic* and *analgesic.*

Salicylate. A salt of salicylic acid.

Salicylic Acid. 2-hydroxybenzoic acid. $OH.C_6H_4COOH$. A white crystalline solid, m.p. 159°C., used as an antiseptic and for *aspirin.*

Saline. Containing *salt,* A 'saline solution' is a solution of salts in *water,* especially one that is *isotonic* with body fluids.

Salinity. Measure of salt content of sea water or brine.

Slinometer. A hydrometer that is used in determining concentration of salt solution by measuring their density.

Saliva. Secretion of salivary glands; spittle.

Salivary Glands. Glands within mouth cavity which secrete saliva.

Sal Soda ($Na_2CO_3.10H_2O$). Sodium carbonate deca-hydrate; washing soda.

Salt. Compound having an acidic and a basic radical.

Salt Common. *Sodium chloride*, NaCl.

Salt Bridge. A tube of *potassium chloride* in the form of a *gel,* used to connect two *half celis* without mixing the *electrolytes.*

Saltcake. *Sodium sulphate*, $Na_2SO_4.10H_2O$.

Salt Effect. *Salting-out.*

Salt Glands. Birds living at sea and far form land, having pair of glands in head which secrete salts to maintain osmoregulation.

Salting -out. *Precipitation* of a dissolved substance by addition of another that lowers its *solubility; e.g., soaps* can be salted-out by common salt from solutions in water.

Saltpetre. Nitre. *Potassium nitrate.*

Salts of Lemon. Potassium quadroxalate, $KH_3C_4O_8.2H_2O$. A white *soluble* poisonous crystalline *salt,* used for removing inkstains.

Sal Volatile. Commercial 'ammonium carbonate'. It is a mixture of ammonium hydrogencarbonate, NH_2HCO_3, ammonium carbamate, $NH_4O.CO.NH_2$, and ammonium carbonate, $(NH_4)_2\ CO_3$.

Samarium. Sm. Element. R.a.m. 150.35. At. No. 62 R.d 7.536, m.p. 1072°C., b.p. 1791°C. A soft silvery *metal* used in *nuclear reactors* as a *neutron* absorber, in *ferromagnetic alloys* (*e.g.* $SmCo_5$), and, in the form of its *oxide*, in some special glasses.

Sand. Hard, granular powder, of granules of impure *silica* SiO_2.

Sandstone. A *sedimentary rock* formed from *sand* or *quartz* particles goined together with *clay*, calcium *carbonate*, and iron oxide:

Sandwich Compound. A *complex* in which an atom of a *transition element* is sandwiched between two substantially planar *hydrocar-*

bon molecules or groups containing *pi bonds*. Typical example is ferrocene.

Saponification. Hydrolysis of an ester with hydroxide to form soap.

Saponification Number. The number of *milligrams* of *potassium hydroxide* required for the complete *saponification* of one gram of the fat or oil.

Saponine. *Glucosides*, Obtained from plants, that form a later with water. Used as foaming agents and *detergents*.

Sapphire. A natural crystalline form of blue, *transparent corundum* (*alumina*, Al_2O_3); the colour is due to traces of cobalt or other *metals*.

Satellites. Bodies that rotate in *orbits* round other bodies of greater *mass* under the influence of their mutual *gravitational field*.

Satellites, Artificial. In 1957 the first man-made artificial *satellite* was launched by the USSR into *orbit* around the *Earth*. Communication' satellites are artificial Earth satellites used for relaying radio, television, and telephone signals around the curved surface of the Earth. 'Passive' satellites merely reflect the transmissions, from their surfaces, while 'active' satellites are equipped to receive and retransmit signals.

Saturated Colour. Pure colour which cannot be contaminated by white colour.

Saturated Compound. Organic compound containing no free valencies.

Saturated Solution. Solution having maximum amount of solute that can be dissolved at a given temperature and pressure.

Saturated Vapour. Vapour existing in equilibrium with its liquid.

Saturated Vapour Pressure. The *pressure* exerted by a *saturated vapour*.

Saturn (Astr). A *planet*, with 17 small *satellites*, that is surrounded by characteristic rings (see *Saturn's rings*). Its *orbit* lies between those of *Jupiter* and *Uranus*.

Saturn's Ring. These are three concentric rings, composed of ice particles or the remains of [illegible]oken-up *satellite*, which are seen round the *planet Saturn*.

Sawtooth Waveform. A waveform in which the shape resembles the teeth of a saw.

SBR. *Styrene-butadiene rubber*.

Scalar Quantity. Any quantity which is sufficiently defined when the magnitude is given in appropriate units.

Scalene. Triangles having three unequal angles and sides.

Scaler. Scaling circuit. An *electronic* device that produces an output *pulse* when a prescribed number of input pulses has been received.

Scandium. (Sc). Elements. R.a.m. 44.956. At. No. 21. R.d 2.99 m.p. 1539°C., b.p. 2850°C. A rare *metal* that occurs in small quantities as the oxide Sc_2O_3 of various isotopes. Only scandium-45, is not radioactive.

Scanning. The repeated and controlled traversing of: a *mosaic* in a television *camera*, or a screen in a *cathode-ray tube*, with an *electron beam.*

Scansorial. Adapted for climbing.

Scattering. The deflection of any *radiation* as a result of its interaction with *matter.*

Scattering of Light. When a *beam of light* traverses a material medium, scattering of the beam takes place. The blue colour of the sky is due to scattering by the actual *molecules* of the *atmosphere.*

Scavenger. Refers to animals that feeds only on dead organism.

Scheel's Green. A bright green *precipitate*, consisting of copper (II) arsenite, $Cu_3(AsO_8)_2$. $2H_2O$. used as a *pigment* and *insecticide.*

Scheelite. An *ore* of tungsten, $CaWO_4$.

Schiff's Reagent. It consists of the dye *magenta*, which has been decoloried with *sulphur dioxide* or *sulphurous acid.* Aldehydes oxidize the reduced form of the dye back to its original colour.

Schmidt Telescope (Camera). Astronomical reflecting *telescope.* The instrument is not used visually but *images* are photographed on a curved surface.

Schrodinger Wave Equation. The *wave equation* used in *wave mechanics* to describe the behaviour of a particle in a field of force, the equation has the form:

$$\nabla^2 \psi + (8 \pi^2 m / h^2) (E - U) \psi = 0$$

where ∇^2 is the *Laplace operator*, ψ is the *wave function*, E is the total energy and U is the potential energy of the particle.

Schweitzer's Reagent. A deep blue solution of a copper ammine in copper (II) hydroxide, Cu $(OH)_2$. A *solvent* for *cellulose*; it was formerly used for this purpose in the obsolete *cuprammonium* process for *rayon* manufacture.

Science. The study of the physical universe and its contents by means of reproducible observations, measurements, and experiments to establish, verify, or modify general laws to explain its nature and behaviour.

Scintillation Counter. A device in which *light* flashes, produced by a scintillator when exposed to *ionizing radiation*, are converted into electrical *pulses* by a *photomultiplier*, thus enabling the number of ionizing events to be counted.

Scintillation Spectrometer. A device for determining the *energy* distribution of a given *radiation.*

Sclerometer. Instrument used to measure the hardness of a material.

Sclerotic. Outer dense coat of vertebrate eye ball.

Scleroprotein. Insoluble, fibrous proteins, occur in surface coatings of animals. They bind cells together in animal tissues.

Scotopic Vision. Vision wherein the *rods* in the eye are the principal receptors. This type of vision occurs when the level of light is low and colours cannot be distinguished.

Screen Grid. Refers to a grid placed between anode and control grid of a thermionic valve.

Scruple. $\frac{1}{24}$ ounce Troy.

Scurvy. Pathological disorder caused due to lack of vitamin C.

Sea-water. Its composition is water, 97.4% common *salt*, NaCl, 2.8% *magnesium chloride*, $MgCl_2$, 0.4%; *magnesium sulphate*, $MgSO_4$, 0.2%; *calcium sulphate*, $CaSO_4$, and *potassium chloride*, KCl, 0.1% each.

Sebaceons. Oil glands connected with hair of mammals.

Sebacic Acid. HOOC $(CH_2)_8$ COOH, decanedioicacid, m.p. 134.5°C, used in manufacture of plasticizer and resins.

Seborrhoea. Refers to excessive discharge from sebaceos glands.

Sebum. Fatty secretion, Lubricates air and skin in mammals.

Second. 1. The *SI unit* of time defined as the duration of 9 192 631 770 *periods* of the *radiation* corresponding to the transition between two hyperfine levels of the *ground state* of the caesium- 133 atom. Symbol s. 2. A measure of *angle*; 1/60 of a minute.

Secondary Cell. *Accumulator.*

Secondary Coil. Secondary winding. The output coil of a transformer or induction coil.

Secondary Colour. A colour, obtained by mixing two *primary colours* *e.g.*, orange.

Secondary Emission of Electrons. When a primary *beam* of rapidly moving *electrons* strikes a *metal* surface. Secondary electrons are emitted from the surface.

Second Derivative (math). The *derivative* of a derivative, written d^2x/dy^2. *e.g., acceleration* (*a*) is the second derivative of *displacement* (*s*) with respect to time (*t*), or the first derivative of *speed* (*v*) with respect to time, *i.e.,*

$$a=d^2s/dt^2=dv/dt.$$

Secretin. Peptide hormone secreted by Brunner's glands in small intestine, it stimulates secretion of pancreatic juice.

Secular Variation of Magnetic Declination. If the *Earth* magnetic North pole is considered to rotate round the geographical North Pole, completing a cycle in about 930 years, a representation of a steady variation of *megnetic declination*, known as the secular, variation, will be seen.

Sedative. A *drug* which reduces nervousness and excitement.

Sedimentary Rock. Rock formed by the accumulation and consolidation of sediments. These include *coal, limestone, sandstone* and *clay.*

Sedimentation. The process of separating an *insoluble solid* from a *liquid* in which it is suspended.

Seed. Multicellular fertilised ovule. It consists of embryo, stored food and seed coat.

Seeding. The addition of fine particles to a *solution* to induce crystallization.

Seger Cone. Pyrometric cone. A device used for estimating the approximate *temperature* of a furnace.

Segregation. To separate paired genes during maturation of sex cells.

Seismograph. An instrument used for recording earthquake shocks.

Seismology. The scientific study of earthquakes and associated phenomenon.

Seismometer. Equipment used for recording earth-quake shocks.

Selenic Acid (H_2SeO_4), Strongly corrosive crystalline acid.

Selenium (Se). Metalloid, exists in four allotropic forms— amporphous bright red, black glassy mass, orange-red crystals and grey crystalline form; used in glass industry and as catalyst, also in photoelectric devices.

Selenium Cell. 1. A *photoelectric cell* consisting of a layer of selenium having a thin covering of gold. Light falling on the cell produces a voltage by the *photovoltaic effect.* **2.** photoelectric cell in which a selenium element changes its *resistance* on exposure to light. This type of cell relies on the *photoconductive effect.*

Selenium Rectifier. A *rectifier* which consists of alternate layers of iron and selenium in contact.

Selenology. The study of the *Moon*, its nature, origin, and movements. Now that samples of the Moon's surface are available for study on Earth, selenology has become a branch of *chemistry* as well as *astronomy.*

Self-absorption. The decrease in the *radiation* from a *radioactive* material caused due to the absorption of a part of the radiation by the material itself.

Self-exciting. (Of a *generator*) having *magnets* that are excited by *current* drawn from the output of the generator.

Semen. Fluid that carries spermatozoa.

Semiconductor Diode. A *semiconductor* device, either based on a *semiconductor junction* or on point contact, with two *electrodes.* Used for rectification.

Semiconductor Junction. A plane that separates two layers of a *semiconductor* each of which have different electrical characteristics.

Semipermeable Membrane. Membrane which permits some molecules, but not all, to pass through it—cellophane or parchment.

Semipolar Bond. Coordinate bond, dative bond. A *velenced bond* in which two *electrons* are donated by one *atom* to another atom, that requires both of them to complete its *octet.* This is equivalent to one electrovalent bond and one covalent bond (see *valency*) and is called a semipolar bond.

Senescence. Refers to phenomenon of ageing or growing old.

Sensitometer. An instrument used for measuring the sensitivity of a photographic plate.

Separation Energy. The *energy* required to remove a particle from a particular atomic *nucleus.*

Sepia. Cuttle fish, ejects a black liquid.

Septavalent. Having a valency of seven.

Septum. Dividing wall between two cavities.

Sequestration. The process of 'locking-up' metal *ions* in *complexes* to make them ineffective.

Series (math). A sequence of numbers such that the *n*th terms may be written down in general form, and any particular term (say, the *r*th) may be obtained by substituting *r* for *n*; *e.g.*, x^n is the general term of the series 1, x, x^2, x^3 ...x^n.

Series, Resistances in. If a number of *conductors* of electricity are connected in series, *i.e.*, one after the other, so that the current flows through each in turn, the total *resistance* is the sum of the separate resistances of the conductors. $R = R_1+R_2+R_3$...

Serine. A white crystalline *amino acid*, m.p. 246°C., occurs in many *proteins*.

Serpek Process. A process for the *fixation of atmospheric nitrogen.* Aluminium is made to react with nitrogen to form aluminium nitride, which is then decomposed by *steam* to give *ammonia.*

Serpentine. A group of magnesium silicate *minerals* with the general formula $Mg_2Si^2O_5(OH)_4$.

Serum. The *liquid* that remains after the clotting and removal of *blood cells* from the *blood*; any similar body liquid.

Servomechanism. A mechanism that converts a small low-powered mechanical motion into a mechanical motion requiring considerably greater power.

Sesame Oil. A yellow oil obtained from sesame seeds, m.p. —6°C., r.d. 0.919, used in the manufacture of *margarine* and cosmetics.

Sessile. Incapable of locomotion ; sedentary ; stalkless.

Sesqui. Indicating that chemical compound contain elements in 2 : 3 ratio.

Shadow Bands. A series of wave *shadow* bands which fall across the *Earth* just before after *totality* in a solar *eclipse.*

Shale. A consolidated laminated *clay rock.*

Shark. Large marine fish. It has tough slate grey skin, attacks man as well.

Shear Stress. A combination of four forces which acting over four sides of a square, give rise to two equal and opposite *couples.*

Shellac. A yellowish natural *resin* secreted by the lac insect (*Laccifer lacca*), it is used in varnishes, polishes, leather dressings and sealing wax.

Shells, Electron. Electrons in the same shell have the same principal *quantum number*, *n*. Each shell is capable of containing $2n^2$ electrons. Within each shell, electrons are further classified into sub- shells (or *energy* sub-levels) according to their orbital *angular momentum*, which is represented by their azimuthal quantum number, *l*. The separate sub-shells are distinguished by the letters s,p,d and f (corresponding to values of *l* of 0, 1, 2 and 3) *e.g.*, and electron designated 4f, has a principal quantum number of 4 (N shell) and an azimuthal quantum number of 3 (f sub-shell).

Sherardizing. Plating iron or *steel* with zinc so as to form a *corrosion* resistant coating.

Shock Wave. Very narrow region of high pressure and temperature wherein air flow changes from subsonic to supersonic.

Shoot. Part of plant axis, generally above ground that bears leaves, buds and flowers.

Short Circuit. If a *potential difference* exists between two points *A* and *B*. If *B* and *A* are joined by a conductor of much lower *resistance* most of the current will flow direct between *A* and *B*, it is then considered as short-circuited.

Short Sight. *Myopia.*

Shower. The production by one high-*energy* particle, originating from *cosmic rays* or *accelerators*, of many fast particles.

Shrimp. Small, long tailed crustacean. It is a valuable food.

Shunt, Electrical. Device used for reducing the amount of *electric current* that flows through a apparatus, such as a *galvanometer*.

Sial Rocks. Forming the Earth's crust below the continents. They are rich in silica and *aluminium.*

Sideband. Frequency band lying on either side of a *modulated carrier wave*; width of each sideband is equal to the highest modulating frequency.

Side Chain (Chem). Refers to an *aliphatic radical* or group that is attached to a *straight chain* or to a *benzene ring* or other cyclic group in the *molecule* of an *organic compound e.g.*, in *methylbenzene* $C_6H_5CH_3$, the *methyl group*, CH_3, is a side chain.

Sidereal Day. The time of a complete rotation of the *Earth* upon its *axis*, with respect to the *fixed stars*. It is 4.09 minutes shorter than a mean *solar day*.

Sidereal Period of a Planet. The 'year' of a *planet*. The actual period' of its revolution round the *Sun*.

Siderite. Iron ore, composed of iron carbonate ($FeCO_3$) ; also called chalybeate or chalybite.

Siderosphere. Iron core at the centre of earth.

Siemen's Process. Process of steel manufacture by addition of scrap metal, iron oxides and ferromanganese to molten pig iron.

Sievert. The *SI unit* of *dose* equivalent. The old unit of dose equivalent, the rem, is equal to 10^{-2} sievert. Symbol Sv.

Sigma Particle Σ-Particle. An *elementary particle* that exists in three charged states: positive, negative and neutral.

Sigma Pile. An assembly of a *neutron* source and a *moderator*, without any *fissile* material. It is used to study the properties of moderators.

Sign Algebraical. The plus or minus sign. + or —.

Significant Figures. The number of digits in a number, as a measure of its accuracy. *i.e.*, the number of digits excluding zeros that indicate only the order of magnitude of the number. *e.g.* 7630 and 0.0675 each contain three significant figures, whereas 6075 contains four.

Silage. A form of cattle-fodder produced by a limited *fermentation* of green fodder pressed down and stored in a pit.

Silane. Silane, silicon hydrides. SiH_4. A colourless *gas*, m.p. —185°C., b.p. - 112°C. Used as a dopant for *semiconductors.*

Silanes. Silicon *hydrides* of the general formula Si_nH_{2n+2}, forming a *homologous series* analogous to the *alkanes*. They are less stable than the alkanes.

Silica. Silicon (IV) oxide, silicon dioxide, SiO_2. A hard *insoluble* white or colourless *solid* with a high *melting point* (1610—1713°C.) Used in the form of a white powder in the manufacture of *glass ceramics* and *abrasives.*

Silica Gel. A form of *silica* SiO_2, having highly porous structure capable of adsorbing 40% of its *weight* of *water* from a *saturated vapour*. Used in *gass* drying and as a *catalyst support.*

Silicates. *Salts* of *silicic acids* various glasses, ceramics, and cements consist largely of silicates.

Silicic Acid. Metasilicic acid, H_2SiO_3 ($SiO_2.H_2O$), and orthosilicic acid, H_3SiO_4 ($SiO_2.2H_2O$), from meta-and orthosilicates.

Silicide. Compounds of silicon with other metal.

Siliciferous. Containing silica.

Silicol Process. Method of preparing hydrogen by reacting sodium hydroxide (NaOH) solution with silicon.

Silicon (Si). Non-metalic most abundant element. Amorphous form is soft, brown, hygroscopic and poor conductor of electricity, dissolves in hydrofluoric acid to produce fluosilicic acid (H_2SiF_6). While crystalline form is hard, grey lustrous and good conductor of electricity, boiling point 2355°C. Used in preparing anticorrosion steel, glass, magnets and as deoxidiser.

Silicon Carbide (SiC). Colourless hard substance used as abrasive for cleaning and polishing metallic surface.

Silicon-controlled Rectifier (SCR). A solid-state electronic apparatus containing three *P-n* junctions, in which the forward anodecathode current is controlled by a third *electrode*, called the *gate*.

Silicon Rubbers. Rubber-like *polymers* of various *organosilicon compound*, such as *siloxanes* highly stable over wide ranges of temperature, good water repellant resistant to chemical action.

Silicones. Refers to *compounds* of the general formula $(R_2SiO)_2$ where *R* stands for *hydrocarbon radicals*. Used as lubricants, for water-repellent finishes, high-*temperature* resisting *resins*, and lacquers.

Siliconising. Diffusing silicon into solid metal at very high temperature.

Silicon Oxide (SiO_2). Exists in amorphous and crystalline forms both insoluble in water acids; used for making glass, cement, synthetic stone etc.

Silicon-steel. Steel containing 0.5 to 4.5% silicon; used for manufacture of electric transformer coils.

Silicon Tetrachloride. $SiCl_4$. Fuming liquid used to prepare silicon compounds and smoke screens.

Silicosis. Disease, caused due to inhaling silica dust. It is a lung disease quite common in miners.

Silk. A thread-like substance produced by the silkworm.

Siloxanes. A group of *compounds* with the general *formula* R_2SiO, where R stands for an *organic* group.

Silver (Ag). White lustrous metal, extermely malleable and ductile, best conductor of heat and electricity, boiling point 2180°C. Found as metal and in minerals aregentite (Ag_2S), horn-silver (AgCl) etc., used for coinage, table ware, jewellery, photography etc.

Silver Bromide (AgBr). Pale yellow precipitate obtained when soluble bromide is added to solution of silver nitrate ; used in photography.

Silver Carbonate (Ag_2CO_3). White precipitate, obtained by adding sodium carbonate to silver nitrate solution or by action of carbon dioxide on silver oxide ; it decomposes on heating into silver, carbon dioxide and oxygen, $2Ag_2CO_3 \rightarrow 4\ Ag + 2CO_2 + O_2$.

Silver Chloride (AgCl). White curdy precipitate formed when hydrochloride acid is added to solution of silver nitrate, in soluble in water but dissolves readily in aqueous ammonia; turns violent in sun light.

Silver Cyanide (AgCN). White curdy precipitate obtained by adding potassium cyanide to silver nitrate solution, not stable compound so changes into metallic silver on addition of water or alkali.

Silver Iodide (AgI). Pale yellow precipitate obtained by adding soluble iodide solution to solution of silver nitrate; used in photography.

Silver Nitrate ($AgNO_3$). Colourless anydrous rhombic crystals, soluble in water, obtained by dissolving silver in dilute nitric acid and evaporating solution to crystallisation ; used in hair dyes, inks, laundry markings etc.

Silver Sulphide (Ag_2S). Black solid obtained by treating silver or its salts and hydrogen sulphide; soluble in alkaline cyanide in presence of air and forms complex alkali argento-cyanide which is used in silver- electroplating.

Silver Thiocyanate (AgCNS). White precipitate obtained by adding potassium thiocyanate to silver nitrate.

Single Bond. Convalent bond between two elements which involves single pair of electrons.

Sintering. Compressing *metal* particles into a coherent *solid* body at *temperature* below its *melting point.* Certain non-metals, such as *ceramics* and *glass*, may also be sintered.

Siphon. Bent-U-tube used to transfer liquid from one container at higher level to another at lower level.

SI Unit. *Systeme international d' unites* is internationally agreed upon system of units for all scientific studies; there are seven basic units-ampere, candela, kelvin, kilogram, metre, mole and second; two supplementary units–radian and steradian.

Sizing. Method of filling pores of paper, that gives some degree of water repellency.

Skatole (C_9H_9N). White soluble crystalline substance having strong odour used in manufacture of perfume.

Skip Distance. The minimum distance at which reception of the sky wave is possible is called the skip distance .

Sky Wave. Refers to radio wave that reaches a point after reflection from ionosphere and not directly from transmitter.

Slag. Glassy material formed by combination of fox and gangue in furance. Solid scum on molten metal.

Slaking. Process of addition of *water.*

Slate. A natural form of aluminium silicate formed from clay and hardened by *pressure.*

Slide Rule. A mathematical instrument used for rapid calculations.

Slow Neutron. A *neutron* with *kinetic energy* not exceeding about 10 *electronvolts.*

Slug. A *unit* of *mass* in the *f.p.s. system* defined 1 slug = 32.174 lbs.

Slurry. A thin paste made up of suspension of a solid in a liquid.

Smelting. The process of chemical *reduction* of the *oxide* of the metal with carbon in a suitable furnace.

Smetic Crystals. *Liqud crystals* having molecules arranged in layers with their axes parallel and perpendicular to the plane of the layers.

Smog. A dark, thick, dust-and soot-laden, *fog*. It pollutes the atmosphere of some industrial cities and the lungs of inhabitants.

Smoke. A suspension of fine particles of a *solid* in a *gas.*

Snell's Law. Radio of sine of angle of incidence to sine of angle of refraction remains constant for any two given media.

Snow. Atmospheric vapour frozen in crystalline form (usually flakes), when air temperature falls below freezing point of water.

Soap. Sodium salt of higher fatty acid ; used for washing and cleaning.

Sodia Lime. Grey solid obtained by adding sodium hydroxide solution to calcium oxide; used as drying agent.

Sodamide ($NaNH_2$). White waxy solid that is obtained by passing dry ammonia over heated sodium metal at 300 to 400°C in iron vessel, $2Na + 2NH_3 \rightarrow 2NaNH_2 + H_2$; used in castner process and in explosive industry.

Soda Water. Water that contain carbon dioxide (CO_2) under pressure releasing pressure lowers solubility of gas and causes effervescence.

Sodium (Na). Silvery-white metal which is prepared by electrolysis of fused sodium chloride, potassium chloride and potassium fluoride mixture, boiling point 883°C. It reacts vigorously with water and halogens ; used as reducing agent, deoxidising agent and as catalyst in polymerisation of isoprene.

Sodium Azide (NaN_3). Colourless crystalline substance that is used in explosives.

Sodium Benzoate (C_6H_5COONa). White powder, dissolves in water; used as food preservative and antiseptic.

Sodium Bicarbonate ($NaHCO_3$). White solid obtained by passing carbon dioxide through cold solution of sodium carbonate, $Na_2CO_3 + CO_2 + H_2O \rightarrow 2NaHCO_3$; used as baking powder and in soft drinks, as medicine to remove acidity of stomach.

Sodium Carbonate (Na_2CO_3). Washing soda; white solid obtained by solvay's process; used in making caustic soda, glass, soap as also in textile and petroleum refining industries.

Sodium Chlorate ($NaClO_3$). Colourless soluble crystalline substance; used as antiseptic, oxidising agent and in explosives.

Sodium Chloride (NaCl). Common salt; white cubic crystals that are anhydrous, main source is sea water; essential constituent of food, used in manufacture of caustic soda, chlorine, sodium carbonate etc.

Sodium Cyanide (NaCN). Colourless solid obtained by heating mixture of crude calcium, cyanamide and sodium carbonate or sodium chloride with coke; used in extraction of gold and silver, in electroplating of copper, gold and silver.

Sodium Cyclamate ($C_6H_{11}NHSO_3Na$). A white crystalline *soluble* powder, used as a sweetening agent in soft drinks but now banned from such use.

Sodium Dichromate (Bichromate). $Na_2Cr_2O_7 . 2H_2O$. An orange *soluble* crystalline substance, m.p. 356.7°C., used as a *mordant corrosion* inhibitor, *oxidizing agent*, and in *electroplating*.

Sodium Ethoxide. Sodium ethylate. C_2H_5ONa. A white *hygroscopic* substance.

Sodium Fluoride. NaF. A colourless, crystalline substance, m.p. 988°C., used in the *fluoridation* of *water* and as an *insecticide.*

Sodium Hydrogencarbonate. Sodium bicarbonate. $NaHCO_3$. A white *soluble* powder, used in *baking power, fire extinguishers*, and in medicine.

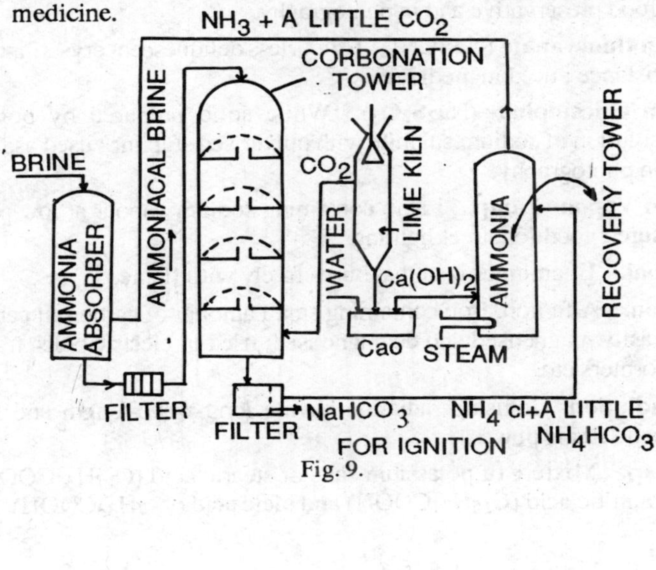

Fig. 9.

Sodium Hydrogenglutamate. Monosodium glutamate. MSG HOOC $(CH_2)_2CH(NH_2)COONa$. A white *soluble* crystalline substance, used to flavour foods.

Sodium Hydroxide. Caustic soda. NaOH. A white *deliquescent solid*, m.p. 318.4°C., dissolves in *water* sed in the manufacture of *soap rayon*, etc.

Sodium Nitrate. $NaNO_3$. A white *soluble* crystalline *salt*, m.p. 306.8°C. It occurs naturally as *Chile saltpetre*. Used as a *fertilizer* and in the manufacture of *nitric acid* and *explosives*.

Sodium Perborate. $NaBO_3$. $4H_2O$. A white *soluble* crystalline substance, m.p. 63°C., used in *bleaching* and as a disinfectant.

Sodium Peroxide. Na_2O_2. A yellow powder, formed on burning sodium *metal*.

Sodium Pyroborate ($Na_2B_4O_710H_2O$). White soluble crystalline salt, found in nature as tincal (crude borax), on heating it loses water of crystallisation and melts to clear transparent solid; used as antiseptic and in manufacture of glass and ceramics.

Sodium Sulphate (Na_2SO_4). White solid, prepared by heating sodium chloride with concentrated sulphuric acid ; used in manufacture of glass and paper.

Sodium Sulphide (Na_2S). Buff coloured solid; used in manufacture of dyes and for removing hair from hides before tanning.

Sodium Sulphite (Na_2SO_3). White crystalline soluble powder; used as food preservative and in photography.

Sodium thiocyanate (NaSCN). Colourless deliquescent crystalline substance ; used in medicine.

Sodium Thiosulphate ($Na_2S_2O_3$). White solid prepared by boiling solution of sodium sulphite with pulverised sulphur ; used as fixer in photography.

Sodium Vapour Lamp. Lamp containing sodium vapour at low pressure ; good for street lighting.

Soft Coal. Bitaminous coal that burns freely with flame.

Soft Iron. Alfa iron. Iron containing small amount of carbon. It can be easily magnetised and demagnetised; used in electric bells, transformers etc.

Soft Radiation. Ionising radiation having long wave-length and low penetrating power.

Soft Soap. Mixture of potassium salts of stearic acid ($C_{17}H_{35}COOH$), Palmitic acid ($C_{15}H_{31}COOH$) and oleic acid ($C_{17}H_{33}COOH$).

Soft Water. Water that lathers with soap.

Soil. The layer of *inorganic rock* material, *organic* matter, and *water* covering most of the *Earth.*

Sol. Colloidal suspension in which liquid is the dispersion medium.

Solar Cell (Battery). An electric *cell* that converts solar *energy* into *electrical energy.* Used in artificial *satellites* and *space probes* to power *electronic* equipment.

Solar Constant. The *energy* received per second by an area of 1 sq metre placed at the mean distance of the *Earth* from the *Sun* and at right angles to the incident *radiation*; approximately 1400 $Js^{-1}\ m^{-2}$ (2 cals $min^{-1}\ cm^{-2}$).

Solar Day. The variable interval between two successive returns of the *Sun* to the *meridian.*

Solar Energy. Energy from sun.

Solar Flares. High temperature outbursts seen as bright areas in the sun's chromosphere. These are associated with sum spots and cause magnetic and ratio disturbances on earth.

Solar Heater. A heater that makes use of solar energy.

Solar Parrallex. Angle between mean equatorial radius of earth as a distance of one astromical unit.

Solar Prominances. Eruption of luminous gas rising some thousand kilometers above the sun's atmosphere.

Solar System. The system consisting of nine planets-*Murcury venus,* the *Earth, Mars, Jupiter, Saturn, Uranus, Neptune and Pluto*— together with their *satellites* revolving in elliptical *orbits* round the *Sun.*

Solar Wind. Streams of electrically charged particles (*protons* and *electrons*) emitted by the *Sun.*

Solder. An *alloy* used for joining *metals.* Soft solders are alloys of tin and lead in different proportions, they melt in the range 200-300°C., brazing solders are composed of copper and zinc, having melting points over 800°C.

Soldering. Use of solder to join two or more metals.

Sole. Lower side of foot.

Solenoid. A coil of wire wound uniformly on a cylinder having a length that is large compared with its radius.

Solid (math). A three-dimensional figure, having length, breadth, and thickness; a figure that occupies *space* or has a measurable *volume.*

Solid Angle. The radio of the area of the suface of the portion of a *sphere* enclosed by the surface forming the angle, to the square of the radius of the sphere.

Solidifying Point. The fixed *temperature* at which a *liquid* soldifies under a given *pressure.*

Solid Solution. A solid *homogeneous mixture* of two or more substances. *e.g.*, some *alloys* are solid solutions of the *metals* in each other.

Solid State. The *physical state of matter* in which the constituent *molecules, atoms*, or *ions* have no *translatory motion* although they vibrate about the fixed positions that they occupy in a *crystal lattice.*

Solid-state Physics. The branch of *physics* that deals with the nature and properties of *matter* in the *solid* state.

Solstice. The points at which the *Sun* reaches its greatest *declination* North or South. The times at which the Sun reaches these points are 21 June and 21 December, the summer and winter solstices, respectively.

Solubility. The amount of a *solute* that will dissolve in a *solvent.* Expressed in kilograms per cubic metre or moles per kilogram of solvent.

Solubility Product. The product of the *concentrations* of the *ions* of a dissolved *electrolyte* when in *equilibrium* with undissolved substance. When the *solubility* product for a given compound is exceeded in a *solution*, it is precipitated out.

Soluble. Capable of being dissolved.

Solute. A substance that is dissolved in a *solvent* to form a *solution.*

Solution. A *homogeneous* molecular *mixture* of two or more substances.

Solvate. Combination of solute and solvent.

Solvation. The combination of *solvent molecules* with *molecules* or *ions* of the *solute.*

Solvay Process. Ammonia-soda process. An industrial preparation of *sodium carbonate* or *wahsing-soda*, Na_2CO_2. $10H_2O$.

Solvent. A substance capable of dissolving other substances in it.

Soma. Animal or plant body excluding germ cells.

Sonar. Refers to sound navigation and ranging; device for locating under-water objects by transmitting high-frequency sound pulse and receiving the reflected pulse ; time taken by the pulse (wave) to reach the object and return gives idea of the depth.

Sonolysis. Process of breaking up of molecules by ultrasonic radiation.

Sorbent. Any agent that is used for *sorption.*

Sorbite. The constituent of *steel* obtained when *martensite* is *tempered* above 450°C., consisting of *ferrite* and finely divided *cementite.*

Sorbitole $CH_2OH(CHOH)_4CH_2OH$. A white crystalline sweet *soluble polyhyric alcohol,* m.p. 110°C., obtained from *dextrose;* used as a *sugar* substitute and in the manufacture of synthetic *resins.*

Sorel's Cement. *Magnesium chloride* cement.

Sorosis. Fleshy fruit such as pineapple. These are formed from cluster of flowers.

Sorption. *Adsorption* or *absorption.* The term is used when the mechanism of a process is not known or is not specified.

Sound. A physiological sensation received by the ear. It is produced by a vibrating source having a *frequency* in the range 20-20 000 *hertz.*

Sound, Speed of. The *speed* of propagation of *sound* waves. In air at 0°C. it is 332 metres per second or approximately 760 miles per hour.

Source. Refers to the *electrode* in a *field-effect transistor* from which electrons or *holes* enter the inter-electrode space.

Soxhlet Extraction Apparatus. An apparatus for extracting the *soluble* portion of any substance by a continuous circulation of the boiling solvent through it.

Space. The part of the four-dimensional *continum* in which *matter* is physically extended. That is capable of travelling in *space.*

Space Probe. Space probes are used for study of conditions within the *solar system* that cannot be studied by terrestrial observation.

Space-time. Based on the theory of *relativity* the disap pearance of a clear-cut distinction between a three-dimensional *space* and an independent time.

Spallation. Nuclear reaction in which a photon or any other high energy particle emits several particles from target nucleus.

Spanaemia. Refers to red blood corpuscles deficiency in blood.

Specific Heat. Quantity of heat required to raise temperature of unit mass (one gram) of substance by 1°C.

Specific Impulse. The ratio of the *thrust* produced to the rate of *fuel* consumption.

Specific Resitance *Resistivity.*

Specific Surface. The total surface area per unit mass of a given substance.

Specific Volume. The *volume* at a specified *temperature* and *pressure*, occupied by unit mass (usually 1 kg) of a substance.

Spectral Series. The emission *spectrum* of the *atoms* or *ions* of a gas when analysed into one or more groups of *frequencies* (or *wovelengths*), the frequencies in each group form a series.

Spectral Types. Spectral classes. The classification of *stars* based on the *spectrum* of the *light* they emit.

Spectrograph 1. An instrument by which *spectra* may be photographed. **2.** A photograph taken by means of such an instrument.

Spectrographic Analysis. An investigation of the chemical nature of a substance by the examination of its *spectrum.*

Spectometer. Type of spectroscope used for precise measurement of refractive indices.

Spectrometer. Apparatus used for analysing spectrum of a source of light or measuring refractive indices.

Spectroscope. Instrument for obtaining and observing spectra.

Spectroscopic Binary. A *binary star* system which cannot be seen as two stars by a *telescope*, but which show a *Doppler effect* in their *line spectrum.*

Spectroscopy. The study of *matter* and *energy* by the use of a *spectroscope.*

Spectrum (Pl. Spectra). An after image; range of colours produced by prism; range of electromagnetic radiation emitted or absorbed by a structure under specified conditions.

Spectrum Colours. The seven *colours* visible in the continous *spectrum* of *white light;*

Specular Reflection. Regular reflection of *electromagnetic radiation.*

Speculum. A reflecting *mirror*, used in a reflecting *telescope.*

Speculum Metal. An *alloy* of 2/3 copper and 1/3 tin ; used for *mirrors* and *reflectors.*

Speed. The ratio of the distance to the time of a moving body.

Spelter. Commercial zinc, about 97% pure, containing lead and other impurities.

Sperm. Male gamete or germ cell; spermatozoa ; seed or semen.

Spermaceti. Waxy solid consisting mostly of cetyl palmitate ($C_{15}H_{31}COOC_{16}H_{33}$), obtained from whale head and used in manufacture of soaps and cosmetics.

Spermary. Male germ glands which produce sperms.

Spermatheca. Refers to serminal receptacle in female insects. The sperms are received and stored here.

Spermatocycte. Male gametocyte which undergoes meiosis and forms spermatids, these spermatids then change into spermatozoa.

Sphenodon. Primitive reptile which was lizard like, found in New zealand.

Sphere (Math). A *solid* figure obtained by the revolution of a semcircle about a diameter as *axis. Volume* of a sphere having radius $r=4\pi r^3/3$; surface area = $4\pi r^2$.

Spherical Coordinates. Three-dimensional *polar coordinates.*

Spherical Triangle. A *triangle* drawn on a spherical surface. The properties of such triangles differ from those of *plane* triangles;

Spherical Trigonometry. *Trigonometry* dealing with *spherical triangles.*

Spheroid. A *solid* figure obtained by an *ellipse* rotating about its minor *axis.*

Spherometer. Instrument for measuring curvature of surface.

Sphygmagraph. Instrument used for recording pulse-beat.

Sphygmology. Science relating to pulse and pulse-beat.

Sphygmomanometer (Sphygmometer). Apparatus used to measure arterial blood pressure.

Sphygmophone. Apparatus by means of which pulse-beat makes sound.

Sphygmoscope. Contrivance with the help of which arterial pulsation becomes visible.

Sphygmus. Pulse.

Spiegel. Alloy of iron, manganese and carbon. Also known as spliegeleisen used in manufacture of steel.

Spinels. Minerals with formula MO, R_2O_3 where M is bivalent metal and R any trivalent metal.

Spiral Galaxies. Spiral nebulae. *Galaxies* in which the *stars*, dust, and *gas* clouds are concentrated in the arms of a spiral.

Spirans. *Compounds* whose *molecules* have two rings sharing a common atom.

Spirillum. A spiral-shaped *becterium.*

Spirits of Salt. A *solution of hydrochloric acid.*

Spirits of Wine. *Ethanol.*

Spiro Compounds. *Spirans.*

Spontaneous Combustion. Combustion of any substance having low ignition temperature and results from the heat formed in the substance due to slow oxidation.

Sputtering. A process for depositing a thin uniform film of a *metal* on to a surface.

Square. 1. A *quadrilateral* with all its sides equal and all its *angles* 90°. **2.** The square of a quantity is quantity raised to the second *power.*

Square Wave. A *wave motion* which alternates between two fixed values for equal lengths of time.

Squaring the Circle. The problem of constructing a *square* exactly equal in area to a given *circle.*

Stabilization. Prevention of chemical decomposition by addition of a stabilizer.

Stable (chem.) Not readily decomposed.

Stable Equilibrium (phys.). A body at rest is in stable *equilibrium* if, when slightly displaced, it tends to return to its original position of equilibrium.

Stainless Steel. *Steel* containing 70%-90% iron, 12%-20% chromium, 0.1% -0.8% carbon. It has many industrial and domestic uses, especially as the alloy 18-8 (18% Cr, 8% Ni, and 0.08% C).

Stalactite. Refers to the downward growth of *calcium carbonate,* $CaCO_3$, formed on the roof of a cave by the trickling of *water* containing calcium *compounds.*

Stalagmite. Refers to the upward growth from the floor of a cave of *calcium carbonate,* $CaCO_3$.

Stalagmometry. The measurement of *surface tension* by determining the *mass* (or *volume*) of a drop of the *liquid* hanging from the end of a tube.

Stallion. Uncastrated male horse used for breeding.

Stalloy. Steel containing 3.5% silicon used in electrical apparatus which are subjected to alternating magnetic fields.

Stamen. Male reproducing organ in flowering plants.

Staminate. Refers to those flowers that have stamens but have no carpels.

Standard Cell. A specially prepared *primary cell, e.g.,* the *Weston cell,* having a constant *E.M.F.,* over long periods of time.

Standard Deviation. It is defined as the *square root* of the average value of the *squares* of the deviations from their mean value.

Standard Electrode. *Hydrogen electrode; calomel electrode.*

Standard Form. A form in which large and small numbers are written so that only one digit appears before the decimal point, and the magnitude is indicated by multiplying by 10 raised to the appropriate power.

Standard Temperature and Pressure. *s.t.p.* 0°C and 760 mm.

Standing Wave. Stationary wave. A wave produced by the simultaneous transmission of two similar *wave motions* in opposite directions.

Stand Oil. A *drying* oil which has been thickened by heating in an inert atmosphere.

Stannate. A *compound* formed when *tin (IV) oxide* reacts with an *alkali.*

Stannic. Containing tin in its + 4 *oxidation state, e.g.*, stannic chloride, tin (IV) chloride, $SnCl_4$.

Stannous. Containing tin in its +2 *oxidation state, e.g.*, stannous chloride, tin (II) chloride, $SnCl_2$.

Stannum. *Tin.*

Stapes. Stirrup-shaped inner, most ossicle of ear.

Starch. Amylum. *Polysaccharides* consisting of chains of *glucose* units arranged in one of two forms: *amylose* and *amylopectin.* Starch is a white tasteless *insoluble* powder which on *hydrolysis* gives first *dextrin* and finally glucose. Starch is stored by plants in the form of granules and occurs in most seeds.

Stars. Heavenly bodies of a similar nature to the *Sun, i.e.*, intensely hot, glowing masses that produce their *energy* by *thermonuclear reactions.*

Stassfurt Deposits. Natural deposits of several *inorganic salts.* A source of potassium and sodium *compounds* in the form of *carnallite*; also of magnesium bromide, $MgBr_2 \cdot 6H_2O$ and *rock-salt.*

Stat. Prefix attached to the name of electrical *units* to indicate the corresponding *electrostatic unit (e.g., statcoulomb).*

Statcoulomb. The *electrostatic unit* of *electric charge* in the *c.g.s. system*. It is equal to 3.3356×10^{-10} *coulomb.*

Statics. A branch of *mechancis*; the mathematical and physical study of the behaviour of *matter* under the action of *forces.*

Stationary States. Permitted energy levels. A transition from one stationary state to another can only occur with the emission or absorpption of energy in the form of *photons*.

Statistical Mechanics. The study of the properties of large assemblies of particles or components in terms of *statistics*.

Statistics. The study of numerical facts or data and their interpretation in mathematical terms, with special reference to the theory of *probability*.

Stator. The fixed part of any *electric motor* or *generator*.

Steady-State Theory. A theory in *cosmology* that postulates that the *Universe* has always existed in a steady state.

Steam. Water in form of gas, moist or vapour.

Steam Engine. Engine that uses steam under pressure to supply mechanical energy.

Steam Iron. Electric iron that releases steam on to material to be pressed.

Steam Point. The *temperature* at which the maximum *vapour pressure* of *water* is equal to standard atmospheric pressure.

Steam Reforming. The process and converting *methane* from *natural gas* into *carbon monoxide* and hydrogen as starting materials for organic synthesis, *i.e.*

$$CH_4 + H_2O \rightarrow 3H_2 + CO.$$

In this reaction the steam is heated to about 900°C and the reaction takes place over a nickel *catalyst*.

Steapsin. Enzyme found in pancreatic juice. It changes fats to fatty acids and glycerol.

Stearic Acid. Colourless fatty acid, used in making candles, soap etc. (octadecanoic acid.).

Stearine. Stearine. A hard white waxy *solid* consisting mainly of *stearic* and *palmitic acids*. It is made by the *saponification* of natural *fats*.

Stearoyl. The univalent radical $CH_3(CH_2)_{16}CO$ — (from stearic acid).

Stellar Evolution. According to current views *stars* evolve during the course of their history. It is believed that they are born from a *condensation* of *gas* (mostly hydrogen) called a *prostar*, which is compressed as a result of the *gravitational field* between the constituents.

Stellite. An *alloy* of cobait (35%-80%), chromium (15%-40%), tungsten (10%- 25%), molybdenum (0%-40%), and iron (0%-5%). It is hard and non-corroding; used for surgical instruments.

Steradian. The supplementary *SI unit* of *solid angle*.

Stere. A metric unit of *volume*; 1 cubic *metre*.

Stereochemistry. Chemistry of the arrangement in *space* of the *atoms* in a *molecule.*

Stereoisomerism. *Isomerism* caused by different arrangement in three-dimensional space of the *atoms* within a *molecule*, resulting in two isomers that are mirror images of each other.

Stereoregular. Having a regular arrangement in space of the atoms and groups within a *molecule.*

Stereoregular Rubbers. A group of synthetic *rubbers* manufactured by a solution *polymerization* process making use of special *catalysts* to control the stereoisomeric regularity of the products. The structure of natural rubber is thus duplicated, can be used for many of the purposes in place of natural rubber.

Stereoscope. An optical device that can give two-dimensional picture the appearance of depth and solidity.

Stereospecific. Particular arrangement in space of the atoms and groups within a *molecule.*

Steric Hindrance (chem.). Retarding of a chemical reaction on account of the arrangement in space of the atoms of the molecules.

Steroids. *Lipids* include *sterols,* the *bile acids*, certain *hormones* and *glucosides* and *vitamin* D.

Sterols. *Cholesterol* and *ergosterol* are examples of Sterols that are present in many living *organisms* in which they play an essential part.

Stethoscope. Instrument used to hear movements of heart and lungs.

Stibine. Antimoney hybride. SbH_3. A poisonous *gas.*

Stibnite. Natural *antimony trisulphide*, Sb_2S_3. Principal *ore* of antimony.

Stibene. 1.2-Diphenylethene. $C_6H_5CH{:}CHC_6H_5$. Colourless *insoluble* substance, m.p. 124°C, used in the manufacture of *dyes.*

Stilbocstrol $(HO.C_6H_4\ CH{:})_2$. White crystalline *organic compound,* m.p. 171°C, used in medicine as an *oestrogen.*

Still. Device for distillation used for distilling alcohol or petroleum.

Stochastic Process. A process having some element of *probability* in its structure.

Stoichiometric. A stoichiometric compound is one is which *elements* are present in the exact proportions represented by its chemical *formula.*

Stoichiometry. The part of *chemistry* dealing with the composition of substances.

Stockes. *c.g.s.* unit of *kinematic viscosity* equal to the *viscosity* of a *fluid* in poise divided by the density in gram per c.c. 1 centistokes = $10^{-6}m^2s^{-1}$.

Stopping Power. The ability of a substance to reduce the *kinetic energy* of a charged particle passing through it.

Stop Watch. It is used for recording small time intervals.

Storage Battery. *Accumulator.*

Storage Ring. A large evacuated *toroidal* ring that form part of some large particle *accelerators.*

Store. A part of the *hardware* of a *computer* wherein information is stored.

S.t.p. Standard temperature and pressure. Formerly N.T.P. (normal temperature and pressure). A pressure of $1.01325\text{-}10^{-5}$ *pascals* (760 mmHg) and a temperature of 273.15 K (0^oC.).

Straight Chain. A *hydrocarbon* molecule in which the carbon atoms are linked together in one long straight chain.

Strain. (phys.). Ratio of the dimensional change to the original or unstrained dimension. It may be a ratio of lengths, areas, or volumes.

Strain Gauge, Electrical. A grid of fine resistance wire attached by a suitable adhesive to the surface under test. The gauge is suitable for measuring strains of the order of 10^{-4} to 10^{-2}.

Strain Hardening. An increase in the *hardness* and *tensile strength* of a *metal*, due to cold working.

Strangeness. Certain *hadrons decay* more slowly than would be expected from the large amounts of *energy* released in the process. These particles, known as strange particles, have been assigned a *quantum number, s,* to account for this strangeness.

Stratified. Arranged in layers.

Stratosphere. Region of atmosphere, lying between 8 to 16 kilometres above earth's surface where temperature does not fall as altitude increases.

Stratum. Layer of cells in living tissue ; bed of sedimentary rock.

Streamline. A line in a *fluid* such that the *tangent* to it at every point is in the direction of the *velocity* of the fluid particle at that point, at the instant under consideration.

Streptomycin. $C_{21}H_{39}N_7O_{12}$. An *antibiotic* obtained by the *Streptomyces* fungus effective against several types of disease *bacteria.*

Stress (phys.). A *force* per unit area.

Stroboscope. An instrument used to view rapidly moving objects with a periodic motion and to view them as if they were at rest.

Strong Acid. An *acid*, that is completely dissociated into *ions* in solution.

Strong Interaction. An interaction that occurs between *hadrons.* It occurs only at very short range (about 10^{-15} metre) and is the force that holds the *nucleons* together in an atomic *nucleus.*

Strontia. *Strontium oxide* or *Strontium hydroxide.*

Strontium. Sr. Element. R.a.m. 87.62. At. No. 38, *reactive metal* resembling calcium. R.d. 2.54, m.p. 769°C., b.p. 1384°C., occurs as *celestine,* $SrSO_4$ and strontianite, $SrCO_3$. Used in fireworks. The *radioisotope* strontium-90 is present in the *fall-out* from nuclear explosions.

Strontium hydroxide. Strontia. $Sr(OH)_2$. A white *deliquescent* crystalline powder, m.p. 375°C., used in *sugar* refining.

Strontium Nitrate. $Sr(NO_3)_2$. A colourless crystalline substance, m.p. 570°C., used in fireworks and flares to impart a bright crimson colour.

Strontium Oxide. Strontia. SrO. A grey *amorphous* powder, m.p. 2430°C., used in the manufacture of strontium *salts.*

Strontium Unit SU. A measure of the *concentration* of strontium-90 in an *organic* medium relative to the concentration of calcium in the same medium. 1 SU=10^{-12} *curie* of strontium-90 per gram of calcium.

Structural Formula. A chemical *formula* that represent the structure of molecule.

Strychnine ($C_{21}H_{22}N_2O_2$). *Alkaloid* occuring in the seeds of *Strychnos nux vomica.* White crystalline substance, slightly *soluble* in *water*; m.p. 284°C. Used in medicine.

Styrene. Phenylethene. $C_5H_5.CH:CH_2$. A colourless *aromatic liquid,* b.p. 146°C., polymerizes to a *thermoplastic* material (see *polystyrene*) used in the manufacture of synthetic *rubber.*

Styrene-Butadiene Rubber (SBR). A widely used, general purpose synthetic *rubber*, which is a similar to natural rubber.

Subatomic. Consisting of particle smaller than, or forming a part of, the *atom.*

Subcritical. A *nuclear reactor* in which the effective *multiplication constant* is less than unity, and in which the nuclear *chain reaction* is not self-sustaining.

Suberic Acid. Octanedioic acid. $HOOC(CH_2)_6COOH$. A white crystalline *dibasic acid*, m.p. 140°C., obtained from *castor oil*, used in the manufacture of *plastics* and *plasticizers*.

Subgiant. A giant *star* having lower absolute *magnitude* than an ordinary giant.

Sublate. The product obtained by ion *flotation*.

Sublimation (chem). The conversion of a *solid* direct into *vapour*, and vice-versa.

Sub-Shell. A concept used in the *Bohr theory* of atomic structure. Each electron *shell* is divided into sub-shells. In any sub-shell all the electrons have the same azimuthal *quantum number*. The sub-shells are designated by the letters *s, p, d, f.*

Subsonic. Moving at, a *speed* less than Mach 1.

Substantive Dyes. *Direct dyes.*

Substituent. *Substitution product.*

Substitution Product. A *compound* obtained by replacing an *atom* or group with another atom or group in *molecule*. The new atom or group is called the 'substituent'.

Substrate. A substance whose reactivity is increased by a specific *enzyme.*

Subtend (math.). Two points, *A* and *B*, are said to subtend the *angle ACB* at the point C.

Substractive Process. The process of producing *colours*-by mixing three different dyes or pigments together.

Succinate. A *salt* of succinic acid.

Succinic Acid. *Butanedioic acid.*

Succinite. *Amber.*

Sucrase. *Invertase.*

Sucroclastic. *Sugar*-splitting ; *enzymes* that have the power of hydrolyzing complex *carbohydrates e.g., invertase.*

Sucrose. Cane-sugar, beet sugar, saccharose. Common 'sugar' of the household. $C_{12}H_{22}O_{11}$. A white sweet crystalline *disaccharide*, m.p. 160°— 186°C., found in sugar cane, sugar beet, etc.

Sugar ($C_{12}H_{22}O_{11}$). Colourless, crystalline substance, sweet in taste, soluble in water, manufactured from juices of various plants like sugarcane (mainly) and sugarbeet; used as sweetening agent.

Sugar Alcohol. Monosaccharide wherein aldehyde group is reduced to alcohol; found mostly in plants (except glycerol).

Sugar of Lead. Lead acetate [Pb $(C_2H_3O_2)_2$] ; White poisonous crystalline substance, sweet in taste ; used for making white lead, in medicines and as mordant in dyeing.

Sulpha Drugs. Group of organic compounds having sulphonamide group (SO_2NH_2) or its derivatives; used in treating many bacterial diseases.

Sulphaguanidine. Sulphonamide used against dysentery etc.

Sulphation. Conversion of a compound into sulphate by oxidation of sulphide.

Sulphofication. Process of oxidation of sulphur and its compounds into sulphates in the soil by bacterial action.

Sulphur(S). Low-melting dull yellow solid brittle, fusible and inflammable, boiling point 444°C, exists in various allotropic forms; used as disinfectant, fungicide and insecticide.

Sulphuration. Process of chemical combination of an element or compound with sulphur.

Sulphur Dioxide (SO_2). Colourless gas having choking smell; used in bleaching fumigating and as refrigerant.

Sulphur Dichloride Dioxide. Sulphuryl chloride. SO_2Cl_2. A colourless *liquid*, b.p. 69.1°C., used as a chlorinating agent.

Sulphur Dyes. Dyes obtained by heating certain *organic* substances with sulphur and sulphides. These dyes are very cheap and are widely used for dyeing industrial fabrics.

Sulphur Hexafluoride (SF_6). Colourless, odourless gas formed when sulphur burns in presence of fluorine, chemically inert to acids, alkalies, heat and water.

Sulphuric Acid (H_2SO_4). Colourless oily liquid. It is commercially manufactured by contact process. In this process sulphur trioxide is treated with concentrated sulphuric acid in absorption tower to obtain oleum ($H_2SO_4.SO_3$) which is diluted with water to desired concentration of sulphuric acid $H_2SO_4.SO_3 + H_2O \rightarrow 2H_2SO_4$; acts as oxidising as also dehydrating agent ; used in manufacture of fertilisers, batteries etc.

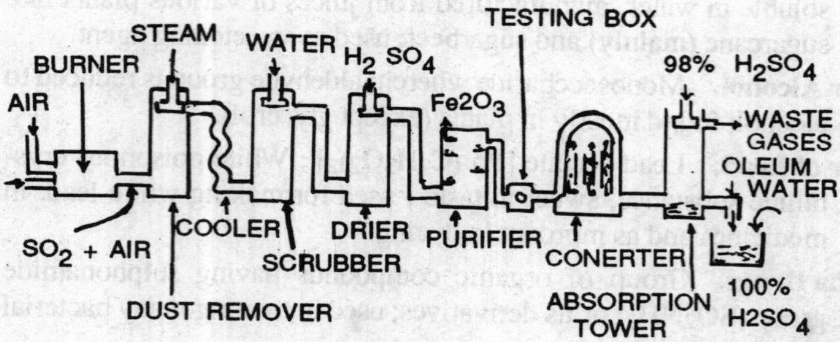

Sulphur Trioxide (SO_3). Fuming gas obtained by heating ferrous sulphate or when mixture of sulphur dioxide and oxygen is passed over platinised silica at 450°C ; used in manufacture of sulphuric acid and for drying gases.

Super Cooling. Cooling of liquid at a given pressure to temperature below melting point without solidifying.

Supercritical. A *nuclear reactor* in which the effective *multiplication factor* axceeds unity, and the nuclear *chain reaction* is self-sustaining.

Superdense Theory. Big-bang theory. The theory that the *Universe* has evolved from one 'superdense' agglomeratian of *matter* due to a cataclylsmic explosion.

Superfluid. A *fluid* capable of flowing without *friction* and having an abnormally high thermal *conductivity, e.g.*, helium-4 below 2.186 K.

Supergiant Star. A *star* of exceptionally high *luminosity*, low *density*, and a very large diameter.

Superheated Steam. *Steam* above 100°C.

Syperheating. Heating a *liquid* above its *boiling point.*

Superpeterodyne. A method of *radio* reception having the *frequency* of the *carrier wave.*

Super High Frequency. S.H.F. *Radio frequencies* in the range 3000 to 30,000 *megahertz.*

Superior Ovary. Calyx, corrolla and stamens attached below ovary.

Supernatant. Refers to a clear *liquid* that floats above a *precipitate.*

Supernovac. *Stars* which suffer an explosion becoming some 10^8 times brighter than the *Sun* during the process.

Superphosphate. · An artificial *fertilizer.*

Superplasticity. The property, shown by certain metallic *alloys,* of stretching several hundred per cent before failing, *e.g.,* zinc in aluminium.

Supersaturation. The state of a *solution* holding more dissolved *solute* than is required to *saturate* the solution.

Supersonic. Moving at a *speed* in excess of Mach 1.

Supersonics. *Ultrasonics.*

Supplementary Angles. *Angles* together totaling 180°, or two right angles.

Suppressor Grid. A grid, placed between the *screen grid* and the *anode* of a *thermionic valve.*

Surd. An irrational quantity ; a *root* that cannot be expressed as an exact number or fraction ; *e.g.,* 1/2.

Surface-active Agent. *Surfactant.*

Surface Colour. Certain reflecting surfaces, *e.g., metal* surfaces, exhibit selective reflection of *light* waves.

Surface Tension γ. It is the force per unit length (*newtons* per *metre*) acting in the surface at right angles to an element of any line drawn in the surface.

Surfactant. Surface-active-agent. A wide-ranging class of substances whose *molecules* contain structurally different groups with opposing solubility tendencies. Depending on their nature, surfactant solutions show as cleansing action (detergency), foaming, wetting, emulsifying, etc.

Susceptance B. The imaginary part of the *admittance* of a circuit. It is measured in *siemens.*

Suspension (chem.). A two-phase system consisting of very small *solid* or *liquid* particles distributed in a *fluid dispersion medium.*

Suspensoid Sol. *Colloidal solutions.*

Sweating. Process of removing entrapped oil and low-melting waxes from filter cake formed in dewaxing processes.

Sylvanicide. Herbicide that kills trees and big plants.

Sylvine (Sylvite). Natural potassium chloride (KCl).

symbiosis. Association between two different individuals, wherein both are mutually benefitted.

Symbol (chem.). A letter or letters representing an *element* or an *atom* of an element; *e.g.* Fe=iron or S=one atom of sulphur.

Symmetry. The correspondence of parts of a figure with reference to a *plane*, line, or point of symmetry. Thus, a *circle* is symmetrical about any diameter.

Syn. Together.

Synandrium. Mass of united stamens.

Synchondrosis. Connection of bones by cartilage.

Synchrocyclotron. A type of *cyclotron* which enables *relativistic* velocities to be achieved.

Synchronous Motor. An *alternating current electric motor* having speed of rotation proportional to the *frequency* of its power supply.

Synchronous Orbit. Stationary orbit. The *orbit* of an artificial Earth *satellite* that has a *period* of 24 hours.

Synchroton. An *accelerator* of the *cyclotron* type whose *magnetic field* is modulated but the *electric field* is maintained at a constant *frequency*.

Synchrotron Radiation. High *energy electrons* within a *synchrotron* emit *light* as a result of their *acceleration* in a strong *magnetic field*; this emission is known as synchrotron radiation.

Synodic Period of a Planet. It is the period between two successive conjunctions with the sun as observed from the earth.

Syntexis. Process of liquefication or melting.

Synthesis. Formation of compound starting from its elements or simple compounds.

Synthetic. Artificially prepared from its component element; not obtained directly from natural sources.

T

Table Salt. Sodium chloride (NaCl).

Tachometer. Aapparatus used for finding out velocity of machines of currents, speeds of aeroplane and motor boats.

Tachycardia. Abnormal rapidity of heart-beat.

Tachyon. A particle that travels faster than the *speed* of light. Its presence could be detected by the *Cerenkov radiation* it emits, but no such particle has yet been detected.

Tachypnoea. Slow but rapid breathing in animals.

Tacnode. The point at which two branches of a curve touch and have a common tangent.

Tacticity[1]. Regularity or symmetry in molecule arrangement in polymer.

Tacticity[2]. Refers to regularity or symmetry in molecular arrangement of structure of polymet molecule.

Tactile. Sense of touch.

Talc [$Mg_3Si_4O_{10}(OH)_2$]. Hydrated magnesium silicate ; extremely soft silvery-white monoclinic mineral, characteristic soapy touch; used as filler, coating and dusting agent, also used in talcum powder and ceramics.

Tall Oil. Resinous substance obtained as by-product in manufacture of wood-pulp ; used in soap and paints.

Tallow. Fat of animal especially cattle. It consists of various glycerides.

Tandem Generator. An *acclerator* of the *electrostatic generator* type.

Tangent Galvanometer. Instrument used for measuring strength of direct current.

Tangoreceptor. Sense organ responsive to touch.

Tannic Acid. White amorphous solid obtained from gall-nuts.

Tannins. *Organic compounds* of vegetable origin. They consist of *mixtures* of *derivatives* of polyhydroxybenzoic acids; *e.g., tannic acid.*

Tantalum Ta. Element. R. a.m. 180.948. At. No. 73. A greyish-white *metal* R.d. 16.6 , m.p. 2996°C., b.p. 5427°C. Occurs together with niobium in a few rare *minerals* used for electric lamp *filaments,* electronic components, in *alloys,* in cemented *carbides* for very hard tools, in surgical components and in electrolytic *rectifiers.*

Tar. Thick, black, sticky liquid got during destructive distillation of coal.

Tartar Emetic. *Antimony potassium tartrate.*

Tartaric Acid. 2, 3-dihydroxybutanedioic acid $COOH.(CH.OH)_2.COOH$. Used in dyeing, calico-printing, and in making *baking-power* and effervescent 'health salts' occurs in grapes.

Tartrate. A *salt* of *tartaric acid.*

Taste. Chemoreception in organisms that enable to discriminate between desirable and undesirable food.

Taste Bud. Collection of gustatory cells and supporting cells, forming organ of taste.

Tau Particle. A heavy *lepton* that reacts by the *weak interaction,* having very short lifetime (less than 1.4×10^{-12} second) and a mass approximately 1782 MeV (*i.e.,* about 3500 times heavier than an electron).

Taurine. $NH_2(CH_2)_2SO_3H$. A white crystalline substance, m.p. 328°C., obtained from the *bile* of mammals.

Tautomerism. Dynamic isomerism. The existence of a *compound* as a mixture of two *isomers* in equilibrium. A substance exhibiting this property is called a 'tautomer'.

Tear Gases. Lachrymators. Substances that can be distributed in the form of a *vapour* or *smoke* and produces an irritating effect on the eyes.

Technetium. Masurium. Tc. Element. At. No. 43. A radioactive metal, m.p. 2172°C , b.p. 4876°C. The most stable *isotope* technetium-97, has a *half-life* of 2.6×10^6 years.

Tektites. Small glass-like bodies having composition unrelated to the geological formations in which they are found.

Telecommunications. The communication of signals , images sounds, or other information by line or *radio* transmission.

Telegraph. A method of transmitting messages over a distance by means of electrical impulses sent through wires. By depressing a key at the transmitting end, a circuit is closed and an electric current flows through the conducting wire or cable to the receiver.

Telemeter. Apparatus used for recording physical events happening at a distance.

Telephone. Apparatus used for conveying sounds to a distant points ; comprises of microphone and receiver.

Teleprinter. Equipment that prints automatically messages sent from one place to another on telegraph lines.

Telescope. Optical instrument used to view distant objects.

Television (TV). Radio transmission of pictures in motion and accompanying sound ; apparatus for receiving such transmissions — black and white as well as colour.

Telluride. Binary compound of any element with tellurium.

Tellurium. To. Element. R. a.m. 127.6 At. No. 52. A silvery-white brittle non-metal, r.d. 6.24, m.p. 452°C., b.p. 1390°C. Exists in several *allotropic forms*, used in *alloys*, for colouring *glass* and in *semiconductors*.

Telstar. Space communication satellite, developed for overseas communication.

Temperature. Degree of hotness or coldness of a body usually measured on a thermometer.

Temperature Coefficient. A constant used to determine variance in physical quantity with temperature.

Tempering Of Steel. Imparting a definite degree of hardness to *steel* by heating to a definite *temperature* and then *quenching i.e.*, cooling, in *oil* or *water*.

Temporary Hardness of Water. Hardness of *water* that is destroyed by *boiling*.

Temporary Magnetism. Induced magnetism. *Magnetism* that a body possess only by virtue of being in a *magnetic field* and that disappears when the body is remained from the field.

Tenorite. A naturally occurring *copper (II) oxide* consisting of small black scales. Found in volcanic regions and in copper veins.

Tensile Strength. Tenacity. It is measured as a *force* per unit area; *e.g. newtons* per square *metre, dynes* per square centimetre; pounds or tons per square inch. That has to be applied to a material to break it.

Tensimeter. A manometer having sealed bulbs attached to each limb, used for measuring *vapour pressure.*

Tensiometer. 1. An apparatus for measuring the *surface tension* of a *liquid.* **2.** An apparatus for measuring the tension in a wire, fibre or beam. **3.** An apparatus for measuring the moisture content of *soil.*

Tensor. A set of *functions* by which the components of a system are transformed from one system of *coordinates* to another; the ratio in which the length of a *vector* is increased.

Tera (T). Prefix to denote 10^{12} times. 1 Tm = 10^{12} metres.

Terbium. Tb. Element. R.a.m. 158.924. At. No. 65. A silvery metal, r.d. 8.25, m.p. 1356°C. Only stable *isotope* is terbium-159.

Terephthalic Acid. Benzene-1, 4-dicarboxylic acid. C_6H_4 $(COOH)_2$. A white insoluble crystalline substance, the *para-isomer* of phthalic acid, that sub-limes (see *sublimation*) without melting above 300°C. Used in the manufacture of *polyesters* ; in particular, *polyethene terephthalate.*

Terminal 1. (phys.). The point at which an electrical connection is made; **2.** An input or output device connected to a *computer.*

Terminal Speed. A body free to move in a resisting medium when acted upon by a constant *force* accelerates until a certain terminal speed is reached and then the *speed* remains constant.

Terminator. The line on the surface of the *Moon*, or a *planet*, separating the dark and light hemispheres.

Ternary Compound. A *compound* consisting of three *elements.* HNO_3 (*nutric acid*).

Ternary Fission. A *nuclear fission* as a result of which a *nucleus* breaks up into three fragments of comparable *mass.*

Terpenes. *Hydrocarbons* occuring in many fragrant *essential oils* of plants. They are colourless *liquids,* having a pleasent smell with general formula $(C_5H_8)n$ and are made up of *isoprene* units.

Terpineol $(C_{10}H_{17}OH)$. Several *isomeric unsaturated alcohols* occuring in *essential oils.* α -terpineol, m.p. 35°C., b.p. 220°C., is used as a *solvent* and in perfumes.

Tarramycin. Oxytetracycline. $C_{22}H_{30}N_2O_{11}$. An *antibiotic* powder obtained from *Streptomyces rimosus* bacteria, used to combat a wide variety of bacterial infections.

Terrestrial. Refers to living on ground.

Terrestrial Guidance. A method of missile or *rocket* guidance.

Terrestrial Telescope. A *telescope* for use on land or sea.

Tertiary Colour. A *colour* obtained by mixing two *secondary colours, e.g.*, brown and grey.

Tervalent. Trivalent. Having a *valence* of three.

Terylene. A *polyester* fibre derived from benzene-1, 4-dicarboxylic acid, $C_6H_4(COOH)_2$ and *ethanediol* by *condensation* and *polymerization* widely used for making fabrics.

Tesla (T). The *SI unit* of *magnetic flux density,* defined as the density of one *weber* of *magnetic flux* per square *metre.*

Tesla Coil. A *transformer* for producing high *voltages* at high *frequencies,* consisting of a coil the primary circuit of which has a small number of turns but includes a spark gap and a fixed capacitor. The secondary winding contains a large number of turns.

Testis. Male organ which produce sperm cells.

Testosterone. $C_{19}H_{28}O_2$. A male sex *hormone* (*androgen*), in pure form it consists of a white *insoluble* crystalline substance, m.p. 155°C. which promotes the development of male characteristics.

Tetra. Denoting four, fourfold.

Tetraarsenic Tetrasulphide. Realgar. As_4S_4. A *insoluble* poisonous powder, m.p. 307°C., used in the manufacture of fireworks.

Tetrachloroethene. Perchloroethylene. $Cl_2C{:}CCl_2$. A colourless non-flammable *liquid,* b.p. 121°C., used as a *solvent* and in dry cleaning.

Tetrachloromethane. Carbon tetrachloride. CCl_4. A heavy colourless *liquid* having a sweetish smell, b.p. 76.8°C formerly used as a *solvent* and *fire extinguisher.*

Tetracyclines. A group of *antibiotics* obtained from *Streptomycies* bacteria effective against a wide range of bacterial infections.

Tetrad. An *element* having a *valence* of four.

Tetradifon ($C_{12}H_6Cl_4O_2S$). Crystalline solid, insoluble in water, stable in alkali media, light and temperature: used as acaricide, has prolonged residual effect.

Tetraethyllead. Lead tetraethyl (IV). $(C_2H_5)_4$ Pb. A colourless oily *liquid,* used to reduce *knocking* in petrol engines.

Tetraethyl Pyrophosphate. TEPP. $(C_2H_5)_4$ P_2O_7. A colourless *hygroscopic liquid,* b.p. 155°C., used as an *insecticide* and rat poison.

Tetrafluoroethene. $CF_2{:}CF_2$. An unsaturated gaseous fluorocarbon, b.p. -76.3°C., polymerizes into a *thermoplastic* material with good electrical insulation properties sold as Teflon, Fluon.

Tetrahedron. A four - faced *solid* figure contained by four *triangles*; a *pyramid* having a triangular base.

Tetranitromethane. $C(NO_2)_4$. A colourless *volatile liquid*, b.p. 126°C., used as an *oxidant* in *rockets.*

Tetravalent. Quadrivalent. Having a *valence* of four.

Tetrode. A *thermionic valve* containing four *electrodes* ; a *cathode*, an *anode* or plate, a *control grid*, and a *screen grid.*

Thalidomide. $C_{13}H_{10}N_2O_4$. A white crystalline substance, used as a *tranquillizer.*

Thallium. Tl. Element R. a.m. 204.37. At. No. 81. A white malleable *metal* resembling lead, r.d. 11.85, m.p. 303.5°C., b.p. 1460°C. used in *alloys*; its *salts* are used as *insecticides* and rat poisons.

Thebaine. $C_{19}H_{21}NO_3$. A white *insoluble* substance, m.p. 193°C., present in *opium.*

Theodolite. Instrument used for the measurement of *angles*, consists of a *telescope* moving along a circular scale graduated in *degrees.*

Theorem. A statement or proposition which is proved by logical reasoning from given facts and justifiable assumptions.

Theory of Games. A mathematical treatment of competitive games, with reference to the strategic and tactical decisions involving conflicting interests and *probabilities.* The theory is extended for use in military and commercial situations.

Therapeutics. Healing ; remedial treatment of diseases.

Therm. A practical unit of quantity of *heat*; 100,000 *British thermal units*, 25,200,000 *calories*, 1,055,06 × 10^8 *joules.*

Thermal Barrier. The limit to the *speed* with which an aircraft or *rocket* can travel in the Earth's *atmosphere.*

Thermal Capacity. *Heat capacity.*

Thermal Cross-Section. A nuclear *cross-section* as measured with *thermal neutrons.*

Thermal Dissociation. See *dissociation.*

Thermal Equilibrium. The state of a system in which there is no net exchange of *heat* between its components.

Thermalize. To bring *neutrons* into *thermal equilibrium* with their surroundings.

Thermal Neutrons. *Neutrons* of very slow speed and of low energy. Their *energy* is the same as the thermal energy of the *atoms* or *molecules* of the substance through which they are passing; i.e. about 0.025 eV.

Thermal Reactor. A *nuclear reactor*, most of its *nuclear fissions* are due to *thermal neutrons.*

Thermal Spike. The zone of high *temperature* briefly produced in a substance along the path of a high *energy* particle or *nuclear fission* fragment.

Thermion. An *ion* emitted by a hot body.

Thermionic Emission. The emission of *electrons* from a heated *metal.*

Thermionics. The branch of *electronics* dealing with the emission of *electrons* from substances at high temperatures.

Thermionic Valve or Tube. A system of *electrodes* arranged in an evacuated *glass* or *metal* envelope.

Thermistor. A *semiconductor*, whose electrical *resistance* decreases rapidly with increase of *temperature;* used as a sensitive temperature-measuring device.

Thermite. Thermit. A *mixture* of aluminium powder and the *oxide* of a *metal, e.g., iron (III) oxide.* The mixture is used for *welding* iron and *steel*, and in incendiary bombs.

Thermobarograph. Instrument for measuring and recording atmospheric *temperature* and *pressure.*

Thermochemistry. The branch of *physical chemistry* that deals with the quantities of *heat* absorbed or evolved during *chemical reactions.*

Thermocouple. Instrument used for measurement of *temperature.*

Thermodynamics Energy. *U. Internal energy.*

Thermodynamics. The study of the general laws that govern processes involving *heat* changes and the conservation of *energy.*

Thermodynamics, Laws of. The law of the conservation of *energy.*

Thermoelectricity. Effects that arise from interaction between currents of heat energy and electrical energy in any materials.

Thermograph. An apparatus that records *temperature* variations during a period of time on a *graph.*

Thermographic Analysis. Methods of chemical analysis based on recording changes of mass or of temperature when substances undergoing chemical changes on heating are heated at a definite rate.

Thermography. A technique used in medicine to detect cancers.

Thermoluminescence. *Luminescence* resulting from a rise in the *temperature* of a body or substance. It occurs if *electrons*, trapped in crystal *defects*, are freed by heating the crystals.

Thermometer. It is an apparatus used for recording temperature.

Thermopile. Instrument for detecting and measuring *heat* radiations.

Thermoplastic. Substance that becomes plastic when heated ; a *plastic* material that can be repeatedly melted or softened by heat without change of properties.

Thermosetting Plastics. *Plastics* that, lose their plasticity on heating.

Thermosphere. The region of the *upper atmosphere* in which the *temperature* increases with *altitude.*

Thermostat. Instrument for maintaining a constant *temperature.*

Thiamine. Aneurin. Vitamin B_1. $C_{12}H_{17}ON_4SCl$. A member of *vitamin* B complex; a white crystalline powder *soluble* in *water* and *alcohol,* Widely required by many living *organisms* for the *metabolism* of *carbohydrates* occurs in liver, milk eggs, and fruits. Deficiency leads to beriberi and nervous disorders.

Thiazines. A group of *compounds* having a six membered ring, four of which are carbon *atoms*, one of which is a sulphur atom, and one a nitrogen atom.

Thiazole. S.CH:N.CH:CH. A colourless *volatile liquid*, b.p, 116.8°C, whose *molecule* consists of a five-membered ring.

Thigmotropism. Response to stimulus of touch.

Thin-Layer Chromatography. A form of *chromatography* in which the stationary phase consists of a thin layer of alumina *slurry* on a glass plate.

Thio. Prefix denoting sulphur.

Thioacetamide. CH_3CSNH_2. A colourless *soluble* crystalline substance, m.p. 115- 16°C., used as a source of *hydrogen sulphide.*

Thiocarbamide. Thiourea.

Thiocyanate. A *salt* of *thiocyanic acid.*

Thiocyanic Acid. HSCN. An *unstable acid,* forms *salts* called *thiocyanates.*

Thio Ethers. A group of compounds with the general formula RSR′, where R and R′ are alkyl radicals.

Thiokols. *Rubber-like polymer* materials with general formula $(Rs_x)_n$ where R is an *organic bivalent radical*, and x is usually between 2 and 4 very resistant to the swelling action of *oils*, and undergo a form of *vulcanization* when heated with certain metallic *oxides.*

Thiolates. Metallic salts of *thiols*, known as "mercaptides"; sulphur analogues of *alcoholates.*

Thiols. A class of organic compounds of the general formula RSH, having sulphur attached directly to carbon.

Thiomin ($C_{12}H_9N_3S$). Dark brown thiazine derivative ; used as stain in microscopy.

Thionin $C_{12}H_9N_3S$. A dark brown *thiazin derivative*, used as a *dye* in microscopy.

Thionyl Chloride. *Sulphur dichloride oxide.*

Thiophene. (C_4H_4S). A colourless liquid *heterocyclic* compound, b.p. 84.0°C., having a nauseating stench occurs in coal-tar ; used as a solvent and in the manufacture of dyes, plastics, and pharmaceutical products.

Thiosulphate. A *salt* of *thiosulphuric acid.*

Thiosulphuric Acid. $H_2S_2O_3$. An *unstable acid* obtained by replacing one oxygen *atom* of *sulphuric acid* (H_2SO_4) by one sulphur atom.

Thiourea. Thiocarbamide. $NH_4CS.NH_2$. A colourless *organic compound*, m.p. 180°C., used in the manufacture of thiourea-aldehyde *plastics.*

Thixotropy. Property possessed by some gels, they liquefy when shaken or stirred and settle again if left undisturbed.

Thomson Effect. Kelvin effect. A *temperature* gradient along a conducting wire giving rise to an *electric potential* gradient along the wire.

Thomson Scattering. The scattering of *photons* of *electromagnetic radiations* by *electrons* according to classical theory.

Thoria. *Thorium dioxide.*

Thorides. Natural *radioisotopes* that occur in the *radioactive series* containing thorium.

Thorite. A mineral consisting of thorium silicate, $ThSiO_4$; used as a source of thorium.

Thorium. Th. Element. R.a.m. 232.038. At. No. 90. A dark grey radioactive metal, r.d. 11.72, m.p. 1750°C., b.p. 4800°C. The most stable *isotope*, thorium-232, has a *half-life* of 1.4×10^{10} years used in alloys and as *nuclear fuel* in some *breeder reactors.*

Thorium Dioxide. Thoria. ThO_2. A white *insoluble* powder, m.p. 3050°C., used in *gas mantles, refractories* and special *glasses.*

Thoron. A *radioisotope* of radon, radon-220, obtained by the *disintegration* of thorium has a *half-life* of 51.5 *seconds.*

Threnardite. A mineral *sodium sulphate*, Na_2SO_4.

Threonine. A colourless *soluble* crystalline *amino acid*, m.p. 230°C. It is essential to the diet of animals.

Threshold. The lowest value of any stimulus, signal, or agency that will produce a specified effect, *e.g. threshold frequency.*

Threshold Frequency. Minimum frequency of *light* falling on a *metal* surface will give rise to the emission of *electrons.*

Thrombin. *Enzyme* formed in the *blood* of vertebrates that acts upon *fibrogen* to form *fibrin* ; essential to the process of blood clotting.

Thrombocytes. *Blood platelets.*

Thrust. The propulsive *force* produced by a *reaction propulsion* motor.

Thulium. Tm. Element. R.a.m. 168.934. At. No. 69. A soft grey *metal,* r.d. 9.33, m.p. 1545°C., b.p. 1950°C. It has one natural *isotope,* thulium-169.

Thymine. 5-Methyluracil. $C_5H_6N_2O_2$. One of the two *pyrimidine* bases which plays a part in the formulation of the *genetic code.*

Thymol. 3-Hydroxy-1-isopropyl-4-methyl benzene. $C_{10}H_{14}O$. A white crystalline *phenol derivative*, m.p. 51.5°C., b.p.233.5°C., smells of thyme, occurs in many *essential oils* and is used as a mild *antiseptic.*

Thyratron. A gas-filled *thermionic valve* where in a *voltage* applied to the *control grid* initiates, the *anode* current. It was formerly widely used as an *electronic* switch but has now been largely replaced by the *thyristor.*

Thyristor. A *silicon-controlled rectifier* wherein the *anode-cathode* current is controlled by a *gate.* It usually consists of three *p-n* junctions on a four-layer chip.

Thyroid. Ductless gland in neck near windpipe that produces hormone (thyroxine); its over activity may lead to exophthalmic goitre. Principal cartilage of larynx forming Adam's apple.

Thyroxin (e). $C_{15}H_{11}I_4NO_4$. An iodine-containing *amino* acid *hormone* produced by the thyroid gland. m.p. 236°C., used in thyroid deficiency cases.

Tides. Movement of the seas caused by the attraction of the *Moon.* At full and new moon the tidal force of the Sun is added to that of the Moon.

Time Exposure. A long *photographic* exposure in which the *camera* shutter is operated manually.

Time-Lapse Photography. Photography in which a slow process, such as the growth of a plant, is photographed by a series of single exposures on cinematic film at regular intervals.

Time Measurement. The *SI unit* of time is the *second*, to which all time-measuring devices are ultimately referred. Such devices include the *pendulum*, the *quartz clock*, the *ammonia clock*, and the *caesium clock*.

Tin (Sn). Soft, easily fusible, malleable and ductile metal not affected by air and water, occurs as white tin and grey tin ; used in tinning utensils and for preparing tin foils.

Tin (Stannic) chloride ($SnCl_4$). Colourless furming liquid, obtained by passing chlorine into molten tin, highly reactive substance.

Tin (Stannic) Oxide (SnO_2). Tin asn ; white insoluble powder, used in getting white glaze for tiles, milky glass articles and electric light shades.

Tin (Stannic) Sulphide (SnS_2). Mosaic gold; yellowish insoluble powder obtained by passing hydrogen sulphide through acidified stannic salt solution; used as pigment for bronzing articles of wood and in manufacture of gold paint.

Tin (Stannous) Chloride ($SnCl_2$). Transparent solid prepared by dissolving tin in hydrochloric acid ; used as reducing agent in laboratory and as mordant in dyeing.

Tin (Stannous) Oxide (SnO.) Dark grey powder that is amphoteric oxide and dissolves in acid and alkali ; burns in air to form stannic oxide.

Tin (Stannous) Sulphide (SnS). Grey crystalline compound obtained by burning tin in sulphur vapour ; on heating above 265°C it turns into stannic sulphide.

Tincture. Alcoholic extract or solution in alcohol.

Tinning. Method of depositing a very thin layer of tin on utensils.

Tintometer. Instrument used for comparing the colour of *solutions*.

Tints. *Colours* having some *hue* but different *saturation*.

Titanium (Ti). Steel grey transition metal, light, ductile, having high tensile strength and low coefficient of expansion, boiling point 3280°C, most abundant ore is titaniferous iron stone ($FeTiO_3$) and rutile (TiO_2); used in constructing aircrafts, ships, marine machinery etc.

Titanium Dioxide (TiO_2). Titania ; white powder insoluble in water, reacts slowly with acids but dissolves easily in alkalies forming titanates; occurs in nature as anatase, rutile etc.; used as white pigment in surface coating as also in ceramics, paper and textile industries.

Titanium Tetrachloride ($TiCl_4$). Colourless, strongly fuming liquid, dissolves in water, boiling point 136.5°C., used in preparation of titanium.

Titration. Volumetric technique in which solution of known concentration is added to solution of unknown concentration until end point is reached.

Tobacco Mosaic Virus. TMV. A simple *virus* widely used in biochemical and biological studies, particularly concerning the transference of the *genetic code.*

Tocopherol ($C_{20}H_{50}N_2$). Yellow insoluble substance (Vitamin E) found in vegetable leaves and wheat-germ, stabilises all membranes by preventing oxidation of unsaturated fatty acids; also prevents sterility in rats.

Toluidinec ($CH_3C_6H_4NH_2$). Aromatic amine (methylphenylamine) exists as ortho, meta and para forms: first two are liquids where as para is white crystalline substance, boiling point 200 to 204°C; used in manufacture of dyes.

Tomentose. Refers to leaf surface densely covered with soft hair.

Tomography. Technique for using X-rays for photographing one specific plane of body for diagnostic purposes.

Tone. Refers to character of quality of sound.

Tonometer. Various instruments that are used for measuring pitch of sound, vapour pressure, blood pressure or pressure within eyeball.

Topaz. A crystalline mineral, $Al_2 (OH,F)_2SiO_4$.

Topology. A branch of *geometry* dealing with the way to which figures are 'connected', rather than with their shape or size.

Toroidal. Having the shape of a toroid.

Torque. A *force, moment* of a force, or system of forces that tends to produce rotation.

Torr. A unit of *pressure* used in the field of high vacuum : it is equivalent to 1 mm of mercury or 133.322 *pascals.*

Torricellian Vacuum. The space, containing mercury *vapour*, that is produced at the top of a column of mercury when a long tube sealed at one end is filled with mercury and inverted in a trough of the metal.

Torsion. 'Twisting' about an axis, produced by the action of two opposing *couples* acting in parallel *planes.*

Torsion Balance. The amount of twist produced is used to measure an applied *force.* In the torsion balance, the force to be measured is applied at right angles to, and at the end of, an arm attached to the wire.

Torus (phys). A 'doughnut' or anchor-ring-shaped *solid* of circular or elliptical cross-section.

Total Internal Reflection. If the incident ray meets the surface at such an angle that the refracted ray must be bent away at an angle of more than 90°, the light cannot emerge at all, and is totally internally reflected.

Totality. The period in a total *eclipse* of the *Sun,* during which the bright surface of the Sun is totally obscured from view on *Earth* by the *Moon.*

Tourmaline. Natural crystalline *minerals,* consisting on *silicates* of various *metals* and containing boron.

Toxaemia. Blood-poisoning.

Toxaphene ($C_{10}H_{10}Cl_8$). Chlorinated camphene containing about 70% chlorine, amber coloured waxy solid, insoluble in water but readily soluble in organic solvents like, benzene, xylene; etc., used as insecticide, phytoxic to most cucurbitaceous plants.

Toxicant. Poison ; chemical exhibiting toxicity.

Toxicology. Science dealing with nature and effects of poisons ; qualitative and quantitative study of injurious effects of chemicals.

Tracer. Isotope of an element which is used to study chemical reaction of physical processes.

Trajectory. The path of a *projectile.*

Tranquillizer. A *drug* that reduce tension and anxiety, without impairing alertness or causing drowsiness.

Transmination. The transfer of an *amino group* from one *compound* to another.

Transcendental (Math). 1. Not capable of being expressed as the *root* of an algebraic *equation* with *rational coefficients,* e.g. π or *e.* 2. Not capable of being expressed by a finite number of algebraic operations, *e.g. sin x.*

Transconductance. The *mutual conductance* between the *control grid* of a *thermionic valve* and its *anode*; expressed in *siemens.*

Transducer. Device that receives waves from one or more media or transmission systems and supplies related waves to one or more other media or transmission systems.

Trans-form. *Trans-isomer.*

Transformation, Constant. *Disintegration constant.*

Transformation, Nuclear. The change of one *nuclide* into another.

Transformer. Device by which an *alternating current* of one *voltage* is changed to another voltage, without alteration in *frequency.*

Transient 1. (math). A *function* whose value tends to *zero* as the independent *variable* tends to *infinity.* **2** (phys.). A short-lived oscillation in a system due to a sudden change of *voltage current*, or load.

Transistor. A *semiconductor* device capable of amplification.

Transition Elements. *Elements* that have chemical properties similar to those of their horizontal neighbours in the *periodic table.* These elements having incomplete inner electron *shells* and show variable *valences* : they occur in the middle of the long periods of the periodic table.

Transition, Nuclear. A change in the configuration of an atomic *nucleus.*

Transition Temperature. Transition point. The *temperature* at which one form of a polymorphous substance changes into another ; the temperature at which both forms can co-exist.

Translatory Motion. A motion involving a non-reciprocation movement of *matter* from one place to another.

Translucent. Permitting the partial passage of *light e.g.* frosted *glass.*

Transmission Coefficient. *Transmittance.*

Transmission Electron Microscope. *Electron microscope.*

Transmittance. Transmission coefficient. T. ratio of the intensity after passing through unit distance of the medium to the original intensity.

Transmitter 1. The equipment used to broadcast *electromagnetic radiation* of *radio frequencies.* **2**. The part of a *telephone* system that converts *sound* waves into *electric currents* or the part of a *telegraph* system that converts mechanical movements into electrical currents.

Transmutation of Elements. Process of changing one chemical *element* into another. Once the aim of *alchemy.* Artificial transmutation by suitable *nuclear reactions* is now common in *nuclear physics.*

Transparent Permitting the complete passage of *light.*

Transponder. *Electronic* equipment used to receive a specific signal and automatically transmit a reply.

Transport Number. Transference number. The proportion of the total *electric current* that is carried by a particular type of *ion.*

Transuranic Elements. *Elements* beyond uranium in the *periodic table; i.e.*, elements of *atomic number* greater than 92.

Transverse. Cross-wise ; in a direction at right angles to the length of the body under consideration.

Transverse Waves. Waves in which the vibration or displacement takes places in a *plane* at right angles to the direction of propagation of the wave ; *e.g., electromagnetic radiation.*

Trapezium. A *quadrilateral* with two of its sides parallel. The area of a trapezium having parallel sides *a* and *b* units in length, and vertical

height *h* units is given by $h\ (a + b)/2$.

Tri. Three.

Triangle. A *plane* figure bounded by three straight lines.

Triangle of Forces. If three *forces* acting at a point can be represented in magnitude and direction by the sides of a *triangle* taken in order, they will be in *equilibrium.*

Triangle of Velocities. If a body has three component *velocities* that can be represented in magnitude and direction by the sides of a *triangle* taken in order, the body will remain at rest.

Triatomic. Having three *atoms* in the *molecule, e.g., ozone,* O_3.

Triazine. $C_3H_3N_3$. Three *isomeric compounds* having three nitrogen and three carbon *atoms* forming a six-membered ring.

Triazole. $C_2H_3N_3$. *Compounds* having three nitrogen and two carbon *atoms* forming a five-membered ring.

Tribasic Acid. An *acid* having three *atoms* of replaceable *hydrogen* in the *molecule, e.g. phosphoric (V) acid.*

Tribology. The study of *friction* and lubrication.

Triboluminescence. The emission of light when certain *crystals* (*e.g. cane-sugar*) are crushed.

Tribromoethanol. CBr_3CH_2OH. A white crystalline powder, m.p. 79-82°C., used as a veterinary *anaesthetic.*

Tribromoethane. Bromoform. $CHBr_3$. A colourless *liquid,* m.p. 8.3°C., b.p. 149.5°C., used in organic synthesis.

Trichloro Ethanol (CCl_3CHO). Chloral ; colourless liquid, boils at 98°C, soluble in water, ethanol and ether ; used as sedative.

2, 2, 2-Trichloroethanediol. Chloral hydrate. $CCl_3CH(OH)_2$. A white crystalline *solid*, m.p. 57°C., b.p. 97.8°C., used as a sedative and in the manufacture of *DDT*.

Trichloroethylene. $CHCl:CCl_2$. A colourless *liquid*, b.p. 87°C., widely used as industrial *solvent*, in dry cleaning, and as an *anaesthetic*.

Trichloromethane. Chloroform. $CHCl_3$. A *volatile* colourless *liquid* having a strong sweet odour, b.p. 61°C., obtained by the *chlorination* of *methane*, used as an *anaesthetic*, and as a *solvent* and in organic synthesis.

Trichloronitromethane. Chloropicrin. CCl_3NO_2. An oily liquid, b.p. 112°C. A highly poisonous and chemically active substance, used in disinfectants and fungicides.

Triethylamine $(C_2H_5)_3N$. Colourless inflammable liquid, used as a solvent.

Trihydric. Containing three *hydroxyl* groups in the *molecule*.

Trihydroxybenzoic Acid. Gallic acid, $C_6H_2\ (OH)_3COOH$. A yellowish crystalline substance, used in *tanning* and the manufacture of *inks*.

Triodomethane. *Iodoform*.

Trillion. 10^{18}, a million million million (British) : 10^{12}, a million million (American).

Trimer. A substance composed of *molecules* having three molecules of a *monomer*.

Trinitrobenzene. TNB $C_6H_3\ (NO_2)_3$. Used as high *explosives*, and having greater power than *TNT*.

2,4,6-Trinitrophenol. *Picric acid*.

Trinitrotoluene. TNT. Methyl-2,4,6- trinitrobenzene. $CH_3C_6H_2(NO_2)_3$. A pale yellow, crystalline *solid*, m.p. 82°C., made by the *nitration* of toluene, widely used high explosive.

Triode. A *thermionic valve* containing three *electrodes* ; and *anode* or plate, a *cathode*, and a *control grid*.

Triolein. Olein. $(C_{17}H_{33}COO)_3$. C_3H_5. A *glyceride* of *oleic acid*, b.p. 235-240°C *liquid* oil that occurs in many natural *fats and oils*.

Triols. Trihydric *alcohols*.

Triose. A *sugar* containing three carbon atoms in the *molecule*.

Trioxoboric (III) Acid. H_3BO_3. *Boric acid*.

Trioxosulphuric (IV) Acid. H_2SO_3.

Tripalmitin. Palmitin. $(C_{15}H_{31}COO)_3$. C_3H_5. A *glyceride of palmitic acid*, m.p. 65.5°C., b.p. 310-320°C. A fat-like substance occuring in palm-oil and many other natural *fats and oils.*

Triple Bond. Covalent bond obtained between two atoms in which three pairs of electrons are shared.

Triple Point. The point at which gas, liquid and solid phases of substance can coexist in equilibrium.

Triptane. Trimethyl butane ; powerful aviation fuel.

Trisaccharides. A group of *sugars* the *molecules* of which consist of three *monosaccharides.*

Tristearin. Stearin $(C_{17}H_{35}COO)_3.C_3H_5$. A *glyceride* of stearic acid (*octadecanoic acid*), m.p. 53.5°C. It is a fat-like substance that occurs in natural *fats.*

Tritiated Compound. A *compound* in which some hydrogen *atoms* have been replaced by *tritium.*

Tritium (T). Radioactive isotope of hydrogen with mass No.3.

Triton. Nucleus of tritium atom, consisting of one proton and two neutrons.

Trochoid. A curve formed by a point on the radius of a circle as it rolls along a straight line.

Trochotron. A multi-*electrode thermionic valve* used as a *scaler.*

Trona. A *double salt* of *sodium carbonate* and *sodium hydrogen-carbonate*, Na_2CO_3. $NaHCO_3.2H_2O$, found in dried lakes.

Troostite. 1. The constitute of *steel* produced on tempering *martensite* below 450°C., consisting of *ferrite* and finely divided *cementite.* **2.** Troostitic pearlite. The constituent of steel produced on the decomposition of *austenite* when cooled at a slower rate than yields martensite and a faster rate than yields *sorbite.*

Tropine. $C_8H_{15}NO$. A white crystalline *hygroscopic soluble* substance m.p. 63°C.

Tropopause. The boundry between the *troposphere* and the *stratosphere.*

Troposphere. The lower part of the Earth's *atmosphere* where *temperature* decreases with height.

Trouton's Rule. The ratio of the *molar latent heat* of vaporization to the *boiling point* in *kelvin* is a constant for all substances.

Troy Weight 1 grain = 0.0648 gram.
20 grains = 1 scruple.

24 grains = 1 pennyweight.

3 scruples = 1 drachm.

8 drachms = 1 ounce Troy = 1.1 ounces avoirdupois.

Trypanocide. Drug that kills trypanosomes.

Trypsin. Enzyme which changes proteins into amino acids.

Tryptophan. A colourless crystalline *amino acid*, m.p. 281—9°C., essential to the diet of animals, occurs in the seeds of some vegetables.

Tungstate. Salt of tangstic acid (H_2WO_4).

Tungsten (W). Wolfram ; a rare silvery-white transition metal, obtained from wolframite [Fe(Mn)WO_4], cuproscheetile ($CuWO_4$) and scheelite ($CaWO_4$), boiling point 5660°C; used to make lamp filaments and high-speed steel.

Tungsten Carbide (WC). Grey powder formed by direct combination of tungsten and carbon at 1600°C. as hard as diamond ; used to make abrasives and tools.

Tungsten Trioxide (WO_3). Yellowish powder, insoluble in water; used in making tungstates.

Tungstic Acid (H_2 WO_4). Hydrated tungsten trioxide, white crystalline powder, loses a molecule of water at 100°C ; used in lamp filaments.

Tuning Fork. A two-pronged *metal* fork that, produces a pure *tone* of constant specified *pitch* when struck, used in *acoustics* and for tuning musical instruments.

Tunnel Diode. Esaki diode. A *semiconductor* having negative *resistance* over a part of its operating range. Used in switching circuits and where low-noise amplification is required up to *frequencies* of about 1000 *megahertz.*

Tunnel Effect. The passage of an *electron* through a narrow *potential* barrier in a *semiconductor*, inspite of the fact that, the electron does not possess sufficient *energy* to surmount the barrier.

Turbine. Any motor having a shaft steadily rotated by the impact or reaction of a current of *steam, air, water,* or other *fluid* upon blades of a wheel.

Turbogenerator. A steam *turbine* coupled to an electric *generator* for the production of *electric power.*

Turbulent Flow. The type of *fluid* flow wherein the motion at any point changes rapidly in direction and magnitude.

Turkey-Red Oil. A mixture of sulphate esters prepared by treatment of *castor oil* with *sulphuric acid*, used in dyeing.

Turpentine. Oil of turpentine. A *liquid* extracted by *distillation* of the *resin* of pine trees. b.p. 155°C-165°C., composed of *pinene* and used as a *solvent.*

Turquoise. Natural basic aluminium phosphate, coloured blue or green due to presence of traces of copper.

Twaddell Scale. A scale used for measuring the *relative density* of *liquids.* Degrees Twaddell = 200 (r.d.-1) ; r.d. = 1 + degrees Twaddell/200.

Tweeter. A *loudspeaker* which reproduces the higher *audio-frequency* sounds, *i.e.* 5—15 *kilohertz.*

Twyer (Tuyere). Pipe with nozzle (brass or copper) through which air is blown into blast furnace.

Tympanum. Middle ear ; ear drum.

Tyndall Effect. Scattering of light by colloidal particles in path of light making a visible beam ; the principle is used in ultramicroscopes.

Type Metal. Alloy of lead (60 to 80%), antiomy (20 to 30%) and tin (5 to 10%); expands on solidifying and gives sharp cast.

Tyrosine. White crystalline amino acid, prepared from most proteins.

Tyvelose. Dideoxy sugar found in bacterial lipopolysaccharides.

U

Uberous. Producing abundant milk.

Udder. External organ of cow, mare etc. that contain mammary glands; dug ; teat.

Udometer. Rainguage ; pulviometer.

Ultimate Stress. Tenacity.

Ultra. Beyond, extreme or excessive.

Ultracentrifuge. Technique used to separate very small particles with the help of high speed centrifuge.

Ultra-High Frequencies. U.H.F. *Radio frequencies* in the range 300 to 3000 *megahertz.*

Ultramarine. An artificial form of *lapis lazuli*, obtained by heating together *clay, sodium sulphate*, carbon and sulphur.

Ultramicroscope. Instrument, for showing presence of particles that are too small to be seen with the oridinary *microscope.*

Ultrasensual. Not in the range or reach of senses.

Ultra Short. Electromagnetic waves with wave lengths less than 10 metres.

Ultrasonic Frequency. A *frequency* more than 20,000 *hertz.*

Ultrasonic Generator. A device for the production of pressure waves of *ultrasonic frequency.*

Ultrasonics. Supersonics. The study of pressure waves having same nature as *sound* waves, but having frequencies above the audible limit. Ultrasonic scanning is widely used in medicine, for examining fetuses that are likely be damaged by *X-rays.*

Ultraviolet Radiation. Electromagnetic radiation with wave-length range between visible light rays and X-rays (4×10^{-7} to 5×10^{-9}).

Umbra. Region of complete shadow.

Ungula. Part of cylinder or cone cut off by a plane that is not parallel to its base.

Uniaxial Crystal. A crystal possessing one optic axis but double refracting.

Unicellular. Having one cell only.

Unicorn. Mythological horse-like animal having a horn growing from centre of its forehead.

Unified Field Theory. A hypothetical theory that describes the electromagnetic and *gravitational fields* in one set of *equations.*

Uniform Motion. Motion having constant velocity.

Unimolecular. Reaction involving only one molecule.

Unit. A quantity or dimension used as a standard of measurement.

Unitary Symmetry. **SU3.** A method of classifying *elementary particles* on the basis of to their properties. SU3 has successfully predicted the existence of particles which have subsequently been found experimentally, *e.g.*, omega-minus. The concept of SU3 has been extended to a larger symmetry group, called SU4, that leads to the concept known as *charm.*

Unit Cell. The smallest unit of a crystal *lattice*. For example, the *body-centred* and face-centred lattices are forms of a cubic unit cell.

Unit Character. Heredity character which behaves as a unit of transmission to off-springs.

Univalent. Having a valency of one.

Universe. Represents the total of all the *matter, energy* and *space* that man is capable of experiencing. The universe is best described in terms of a four-dimensional curved *space-time continum* ; it contains some 10^{41} kilograme of matter, collected in some 10^{9} galaxies.

Unsaturated Compound. Organic compound containing double or triple bond between two carbon atoms.

Unstable. That can be easily decomposed.

Upper Atmosphere. It usually include gaseous envelope from 30 *kilometres* upwards.

Uracil. Pyrimidinedione. $C_4H_4N_2O_2$. A white crystalline *pyrimidine* base, m.p. 338°C., occurs in the *nucleotides* of *ribanucleic acid.*

Uraninite. *Pitchblende.*

Uranium (U). Silvery-white radioactive metal that exists as U_3O_8 and boils at 3818°C. In India, it occurs in monazite sand in Travancore (Kerala): used to get atomic energy by nuclear fission and in some alloys as source of transuranic elements.

Uranium Dioxide. Black radioactive substances used as fuel in gas cooled reactors.

Uranium Hexafluoride (UF_6). Crystalline volatile compound obtained by heating uranium carbide or pentachloride with fluorine; it is deliquescent and fumes in moist air.

Uranium-Lead Dating. Method of *dating rocks*, it depend on the decay of one of the *isotopes* of uranium to lead. Method based on measuring the ratio of the helium present in a rock sample to the amount of uranium.

Uranium Pentachloride (UCl_5). Deep red deliquescent solid prepared by prolonged chlorination of uranium oxide ; soluble in water and many organic solvents ; dissociates on heating into tetrachloride and clorine $2UCl_5 \rightarrow 2UCl_4 + Cl_2$.

Uranium Trioxide. Uranyl (VI) oxide, UO_3. Red, insoluble, radioactive powder. Decomposes on heating.

Uranus. (astr.) A *planet* having five *satellites*, its *orbit* lying between those of *Saturn* and *Neptune*.

Uranyl. The *bivalent* group, = UO_2, forms *salts* with *acids*.

Urea (NH_2CONH_2). White crystalline solid, soluble in water and ethanol ; used as fertiliser.

Urea Cycle. Method of urea formation in ureotelic vertebrates ; ornithine cycle.

Urea-Formaldehyde Resins. *Thermosetting resins* having good *oil* resistant properties, produced by *polymerization* of *urea* and formaldehyde.

Urease. An *enzyme* capable of splitting *urea* into *ammonia* and *water*.

Ureido. The *univalent radical*, NH_2CONH-.

Urethane Resins. Polyurethanes, *polymers* chemically related to *urethanes*, made by condensation of *isocyanates* with *polyhydric* compounds, used for a number of purposes, such as the manufacture of coatings and foam plastics.

Urethan(e)s. *Esters* of carbamic acid, NH_2COOH. The name is usually applied to ethyl carbamate, $NH_2COOC_2H_5$.

Ureylene. Divalent radical-NHCONH -, derived from urea.

Uric Acid ($C_5H_4O_3N_4$). Nitrogenous waste product found in urine of aves, reptiles etc.

Urochrome. Pigment that imparts yellow colour to urine.

Urokinase. Enzyme found in urine of human beings, catalyses conversion of plasminogen to plasmin.

Urology. Branch of medical science that involves study of urine.

Uroscopy. Examination of urine for diagnostic purposes.

Uveitis. Inflammation of iris, ciliary body and choroid.

V

Vacancy. Schottky defect, irregularity that occurs in a *crystal lattice* when a site normally occupied by an *atom* or *ion* remains unoccupied.

Vacuum. A *space* having no *molecules* or *atoms.*

Vacuum Distillation. The process of *distillation* at a reduced *pressure.* This process enables those substances to be distilled, which would decompose at normal pressures.

Vacuum Evaporation. The substance is heated in a *vacuum,* the atoms that escape from its surface are allowed to condense on the surface to be coated.

Vacuum Pump. A device used to produce a low pressure.

Valence. Capacity of atom to combine with or to be replaced by another atom.

Valency. Unit of valence-carbon has four valencies.

Valence Band. The range of energies in a *semiconductor* that correspond to states that can be occupied by the *valence electrons* binding the *crystal* together.

Valence Bond. The link that holds *atoms* together in a *molecule.* Two *univalent* atoms are held together by a single valence bond ; it is possible for an atom to satisfy two or three valence bonds of another atom that gives rise to a *double* or *triple bond.*

Valence Electron. An outer *electron* of an *atom* taking part in formation of a *valence bond,* electrovalent.

Valeric Acid. *Pentanoic acid.*

Valine. White crystalline, essential amino acid with unsubstituted chain; found in most proteins.

Vanadate. A *salt* of *vanadic acid.*

Vanadic Acid. HVO_3. A yellow insoluble crystalline substance. Other *acids* are formed by the addition of *water molecules, e.g.* H_3VO_4.

Vanadium (V). Very hard and tenacious metal that boils at 3388°C ; used in making vanadium steels and its compounds.

Vanadium Pentaoxide (V_2O_5). Orange-red coloured powder that melts at 600°C and recrystallises on cooling.

Vanadyl. The *divalent* group = VO_2.

Van Allen Radiation Belts. Two belts of charged particles trapped within the Earth's *magnetic field.*

Van De Graaff Generator. An *electrostatic generator* for producing high voltages for use with particle accelerators and for other purposes.

Van Der Waals' Equation of State. For a mole of a real gas

$$(p+a/V^2)(V-b) = RT,$$

where p = *pressure*, V = *volume*, T = *thermodynamic temperature*, R = *the gas constant.*

Van Der Waals' Force. An attractive *force* that exist between *atoms* or *molecules* of all substances. In many substances this force is small compared with the other inter-atomic attractive and repulsive forces present.

Vanillin. $CH_3O(OH)C_6H_3CHO$. A white *insoluble* crystalline substance, m.p. 80-81°C., obtained from vanilla beans used as a flavour and in perfumes.

Van't Hoff's Law. The *osmotic pressure* of a dilute *solution* is equal to the *pressure* that the *solute* would exert in the gaseous state, if it occupied a *volume* equal to the volume of the solution, at the same *temperature.*

Vapour. A substance in the gaseous state which can be liquefied by increasing the *pressure* without altering the *temperature.* A *gas* below its *critical temperature.*

Vapour Density. *Density* of a *gas* or *vapour* ; usually as relative to hydrogen. The ratio of the *mass* of a certain *volume* of the gas to the *mass* of an equal volume of hydrogen, measured under the same conditions of *temperature* and *pressure.* Numerically this ratio is equal to half the *molecular mass* of the gas.

Vapourisation. Process by which liquid or solid is converted by heat into gas or vapour.

Vapour Pressure. Pressure of vapour in equilibrium with the liquid or solid.

Varec. Kelp. The *ash* of seaweed, from which iodine is extracted.

Variable (math.) A symbol or term that assumes, different numerical values, *e.g.*, in $y = 5x^2 + 2$, x is the independent variable and y is the dependent variable.

Variance. **1** (statistics). The *square* of the mean *deviation.*
2. (chem.). The number of *degrees of freedom* that a system can have.

Variate. (statistics) A *variable* that can have any of a *set* of values according to specified *probabilities.*

Variation. (math.) If a quantity y is some *function* of another quantity x, *i.e.*, if $y=f(x)$, then, as x varies, y varies in a manner determined by the function.

Variometer. A *variable inductance* consisting of two coils in *series,* arranged so that one coil can rotate within the other. Used as a means of measuring inductance.

Varicella. Chickenpox.

Variola. Smallpox.

Vaseline. Petroleum jelly (*Paraffinium molle*) and its preparations.

Vasomotor. It serves to regulate tension of blood-vessels.

Vat Dyes. A class of *insoluble dyes* that are applied by first reducing them to leuco-compounds, *soluble* in *alkalis. Indigo* and many synthetic dyes belong to this class.

VDU. *Visual-display unit.*

Vectors. A physical quantity that requires a direction to be stated in order to define it completely, *e.g. velocity.*

Vectors, Parallelogram of. A particle under the action of two like *vector* quantities, are represented by the two sides of a *parallelogram* drawn from a point, the *resultant* of the two vectors is represented in magnitude and direction by the diagonal of the parallelogram drawn through the same point.

Vectors, Triangle of. A particle acted upon by two *vector* quantites represented by two sides of a *triangle* taken in order, the *resultant* vector is represented by the third side of the triangle.

Vector. Agent or carrier that transmits bacterial, fungal or viral pathogens from one host to another.

Vegetable Oil. Oil obtained from various plant parts (leaves, stems or seeds); consists of esters of fatty acids and glycerol.

Vegetative Nucleus. Large nucleus formed within pollen grain of Angiosperms along with one or two smaller generative nuclei.

Velocity. Swiftness or speed in given direction; distance covered by a body per unit time in given direction.

Velocity, Relative. It is the rate at which a body is changing its position with respect to the other. If the velocities of two bodies are represented by two sides of a *triangle* taken in order, thier relative velocity is represented by the third side.

Velocity Modulation. The *modulation* of the *velocity* of a stream of *electrons* by alternately accelerating and decelerating them.

Velocity of Light. *Speed of* light.

Valocity Ratio of a Machine. Distance ratio. The ratio of the distance through which the point of application of the applied *force* moves, to the distance through which the point of application of the resistance moves in the same time.

Venetian White. A mixture of lead *carbonate* (white lead) and *barium sulphate* (blanc fixe) in equal parts, used in *paints.*

Venturi Tube. A device for measuring the rate of flow of a *fluid.*

Venus. (astr.) A *planet* with its *orbit* between those of *Mercury* and the *Earth.*

Verdigris. Copper (cupric) acetate or the green coating of cupric carbonate which forms on copper, brass or bronze in presence of atmosphere.

Vermicide. Any substance used to kill worms.

Vermiculite. A group of low-grade *micas* that expand on heating to a light water-absorbent material, used in the exfoliated form as *heat* and *sound* insulating materials, and in special *soils.*

Vermiform Appendix. Atrophied end of caecum in primates.

Vermilion. A scarlet form of *mercy* (II) *sulphide,* HgS ; used as a *pigment.*

Vernier. A device used for measuring subdivisions of a scale . A vernier consists of a scale that slides alongside the main scale, and on which a length subdivided into equal parts.

Vernier Motor (Engine). A small *rocket* motor used to correct the flight path or *velocity* of a missile or *spacecraft.*

Versed Sine. One minus the cosine of an angle.

Vertex 1. (math.). The point on a geometrical figure farthest from the *base.* **2.** (astr.). The point on the *celestial sphere* towards which, a *star* appears to move.

Very High Frequencies (VHF). *Radio frequencies* in the range 30 to 300 *megahertz.*

Very Low Frequencies (VLF). *Radio frequencies* below 30 *kilohertz.*

Vibration. Regularly repeated to-and-fro motion about a mean position.

Vicinal. When two similar *substitutents* are added to a carbon compound, these are referred to as 'vicinal' (or *vic-*) if the substituents are attached to adjacent carbons.

Video. Pertaining to picture signals ; used in radar or televisions system that carry these signals in modulated or unmodulated form.

Video Cassette (Video Tape). Magnetic tape used for recording television picture and sound; comes in compact case (cartridge) to facilitate its playback.

Video Cassette Player (VCP). Electron machine used for playing pre-recorded video tapes. It is attached to television set for production of picture and sound.

Video Cassette Recorder (VCR). Electron machine, like VCP to playback pre-recorded video cassette on television ; in addition it can also record, on Video tape the programme from TV set-both picture and sound; also called video tape recorder (VTR).

Video Frequency Signal. The signal that transmits the picture and synchronizing information in a *television* system.

Video Tube. Television tube or screen.

Villous. Shaggy appearance due to heavy coverage of long, soft, curly trichomes or villi. Leaf surface thickly covered with long soft hair.

Vinaceous. Having wine colour.

Vinasse. Liquid residue left after fermentation and distillation of beet root molasses. Used as a source of potassium carbonate.

Vinegar. Dilute glacial acetic acid ($HC_2H_3O_2$), containing 3 to 6% ethanoic (acetic) acid, miscible with water, obtained by oxidation of ethanol (C_2H_5OH) by action of bacteria on wine or fermented wort; used as condiment and pickling.

Vinolent. Wine addict.

Vinordinaire. Ordinary red wine, cheap claret.

Vinyl. Unsaturated univalent radical $CH_2.CH^-$.

Vinylene Bivalent radical,—CH :CH—.

Vinyl Ether (C$_4$H$_6$O). Highly inflammable liquid obtained by action of alkali on ethyl ether ; used as anaesthetic.

Vinylidene Chloride (CH$_2$: CCl$_2$). Colourless inflammable liquid that polymerises to form polyvinylidene chloride boiling at 32°C.

Virgin Neutrons. Neutrons produced by any means but before they have experienced a collision.

Viricide. Substance that can destroy or eliminate virus.

Virology. Scientific study of viruses.

Virose. Poisonous, foul.

Virosis. Any diseases caused by virus.

Virulence. Capacity of pathogen to cause diseases ; pathogenicity.

Virulent. Highly poisonous; malignant ; venomous ; acrimonious.

Virus. Ultramicroscopic pathogenic particle having power of growth and reproduction; considered to be transitional stage between living and non-living.

Viscometer. Instrument used for measuring viscosity or viscous flow.

Viscose. A *liquid*, consisting of a *solution* of cellulose *xanthate* in dilute *sodium hydroxide*. Made from *cellulose* by the action of sodium hydroxide and *carbon disulphide*. Used for the production of viscose *rayon* and of cellulose film.

Viscose Rayon. *Rayon.*

Viscosity η. The property of a *fluid* that enables it to resist flowing when it is subject to a shear stress. Viscosity is measured in *pascal seconds* (*SI units*) or *poise* (*c.g.s. units*). 1 centipoise = 10^3 Pas.

Viscous. Resistance to flow due to forces acting between molecules.

Vision. Perception of image derived from nervous impulses originating in eye.

Visive. Able to see; able to be seen.

Visual. Relating to sight.

Visual Agnosial. Inability to see objects by a particular visual modality even though the modality is intact.

Visual Binary. A *binary star* system that can be resolved into two *stars* with an optical *telescope.*

Visual-Display Unit. VDU. A *computer* device whose output is a *cathode-ray* tube for displaying text or diagrams. It may have an input device consisting of a keyboard or it may be a *light pen.*

Visual Threshold. Minimum amount of light that produces sensation of light.

Vital Air. Oxygen.

Vital Capacity. Total quantity of air that can be expelled from lungs —in man, it varies between 3400 and 4000 cubic centimetres.

Vital Force. Force on which phenomenon of life in animals and plants depend.

Vitamin. Complex organic compound, present in natural products or made synthetically; essentially required in small quantities by all animals for their normal metabolism, usually supplied in their diet; absence or shortage causes various deficiency diseases.

Vitamin A. Retinol and Carotene ; found in butter, egg, liver, kidney, milk, vegetables etc., keeps skin healthy, promotes growth and development, improves night vision and resists infections ; deficiency causes night blindness.

Vitamin B-Complex. Group of unrelated water-soluble substances including, vitamins B_1, B_2, B_6 and B_{12}.

Vitamin B_1. Thiamine ; found in egg, fruits, green pea, liver, kidney, potato etc., increases apetite, helps digestion, prevents nervousness; deficiency leads to beriberi.

Vitamin B_2. Riboflavin; found in egg, leafy vegetables, milk etc., keeps skin and muscles healthy, oxidises food and promotes growth ; deficiency may cause glossitis and cheilosis.

Vitamin B_6. Exists in three chemically related and water-soluble forms —pyridoxol, aldehyde pyridoxol and amine pyridoxamine ; found in leafy vegetables and liver ; deficiency may cause convulsions, nervousness and insomnia.

Vitamin B_{12}. Group of closely related polypyrrole compounds containing Cobalt (Co III), found in liver, kidney, meat etc., essential for hemopoiesis ; deficiency may cause, pellagra, characterised by skin eruptions.

Vitamin B_{14}. Not an established vitamin, found in human urine, it increases rate of cell-proliferation in bone-marrow culture.

Vitamin C. Ascorbic acid ($C_6H_8O_6$), found in berries, cabbage, citrus, leafy vegetables, tomato etc., keeps bones, teeth and gums healthy, regulates heart muscles, resists infection ; deficiency may cause scurvy.

Vitamin D. Found in buttercream, egg yolk, fish liver-oils, irradiated milk, sunlight etc., helps in using calcium and phosphorus in bones and teeth ; deficiency causes rickets.

Vitamin E. Tocopherol; found in diary products, meat etc., severe deficiency causes destruction of red blood corpuscles and anaemia.

Vitamin K. Phylloquinone, menaquinone ; fat-soluble vitamin found in egg yolk, green vegetables, milk and milk products; promotes blood- clotting and prevents excess bleeding.

Vitamin L. Not an established vitamin, found in yeast, essential for loctation.

Vitamin T. Mixture of folic acid, niacin and deoxyribosides, found in insect cuticle, mould mycelia and yeast fermentation liquor.

Vitreosil. Silica SiO_2, used for making scientific apparatus which requires to withstand large and sudden temperature changes.

Vitreous. Resembling glass.

Vitreous Humour. Gelatinous fluid behind lens of human eye.

Vitriol. Concentrated *sulphuric acid*, H_2SO_4, oil is vitriol ; *copper (II) sulphate*, $CuSO_4.5H_2O$, called blue vitriol ; *ferrous sulphate*, $FeSO_4.7H_2O$, called green vitriol ; *zinc sulphate*, $ZnSO_4$. $7H_2O$, called white vitriol.

Volcanic Rock. *Igneous rock.*

Volcano. A gap in the Earth's crust from which hot *gases, liquid magma*, and *ash* can escape.

Volt (V). Unit of electrical force that would carry one ampere of current against one ohm resistance.

Voltage. Electrical force measured in volts.

Voltage Divider. Potential divider, potentiometer. A *resistor* or series of resistors connected across a source of *voltage* (*V*) and tapped at a point to give a fraction (*V*) of the total voltage.

Voltage Doubler. An *electronic circuit* that delivers a *direct current voltage* nearly twice the peak *alternating current* voltage it feeds on. It usually consists of two *rectifiers* connected in *series*.

Voltaic Pile. The earliest electric *battery* consisting of a number of *cells* joined in series, each consisting of a sheet of zinc and copper separated by a piece of cloth moistened with dilute *sulphuric acid*.

Voltameter. Electrolytic cell in which a metal, generally silver or copper, is deposited by electrolysis of a salt of metal upon cathode.

Voltmeter. Instrument used for measuring potential difference between any two points in electric circuit.

Volume. Measure of space occupied by solid, liquid or gas.

Volumeter. Instrument used for measuring volume of gas.

Volumetric Analysis. Methods of quantitative chemical analysis involving measurements of volumes of reacting substances.

Volumometer. Apparatus for measuring volume of solid body by quantity of liquid displaced by it.

Vulcanisation. Process of improving quality of rubber by heating it with sulphur.

Vulcanite. Hard insulating material formed by action of rubber on sulphur.

Vulcanized Rubber. Rubber obtained on heating rubber on sulphur.

Vulcan Power. Highly explosive mixture containing 52.5% sodium nitrate, 30% nitroglycerin, 10.5% charcoal and 7.0% sulphur.

Vulgar Fraction. Common fraction. A fraction expressed in terms of a *numerator* and a *denominator e.g.* $^3/_4$.

Vycor Glass. Pure silica glass made from sodium borosilicate glass without production problems of fused silica.

W

Wacker Process. Refers to oxidation of ethene to ethanol cupric chloride in aqueous solution.

Wall Pressure. It is the force exerted by cell wall on cell contents, equal and opposite to turgour pressure.

Wankel Rotary Engine. Epitrochoidal engine. An *internal-combustion engine* using a 4- stroke cycle, but without reciprocating parts.

Warfarin ($C_{19}H_{16}O_4$). Coumarin derivate, colourless crystalline substance, anticoagulant which acts as slow poison; used as rat poison.

Warning (Aposematic) Colouration. Conspicuous markings on animal body (or wings) for defending against predators.

Warning Odour. Distinctive smell imparted to fuel gases for safety purpose; as these gases have little or no odour of their own.

Washing Soda. Na_2CO_3 (crystalline).

Water (H_2O). Clear transparent, colourless, odourless, tasteless liquid, F.P. = O^oC. and B.P. = 100^oC.

Water Barometer. Barometer in which water is used in place of mercury.

Water Gas. Fuel gas consisting of carbon monoxide and hydrogen (1:1) obtained by action of steam on red hot coke ; used as fuel and in industrial preparation of hydrogen gas.

Water of Crystallisation. Water chemically combined in many crystallised substances ; can be removed by heating, results in loss of crystalline properties.

Water Pollution. Contamination of water by certain chemicals, detergents, fertiliser run-off, sewage effluent etc.

Water Repellent. Chemicals used for treating leather, paper, textiles, wood etc., to make these resistant to wetting by water, *e.g.*, various types of resins, aluminium and zirconium acetates.

Water Scorpion. Insect that belong to family Nepidae (Hemiptera), predaceous aquatic bug, forelegs raptorial, hind legs adapted for walking; its bite is painful.

Water Softening. The process of removal of hardness of water.

Water Stress. Drought.

Water Vapour. Water in gaseous or vapour state ; present in atmosphere in varying amounts.

Watt (W). Sl Unit of power, equal to one joule per second.

Wattage. Power measured in watts.

Watt Hour. Unit of energy equal to energy consumed @ one watt in one hour (3600 joules).

Wattmeter. Apparatus used for direct measurement of power of electrical circuit in watts.

Watt Second. Unit of energy equivalent to one joule.

Wave. Propagation of disturbance that transfers energy from one location to another.

Wave Equation. The *equation* that gives mathematical expression to *wave motion* :

$$\nabla^2\psi=1/c^2.\delta^2\psi/\delta t^2$$

where Δ^2 is the *Laplace operator*, ψ is the *wave function*, *c* is the *speed of light*, and *t* is the time at any instant.

Wave Form. The shape of a *wave*, shown graphically by plotting the values of the periodic quantity against time.

Wave Front. The locus of adjacent points in the path of a *wave motion* having the same *phase*.

Wave Guide. A hollow *metal conductor* through which *microwaves* may be propagated. Used extensively in *radar*.

Wave-Length (λ). Distance between two successive similar points on an alternating wave.

Wave Mechanics. A development of *quantum mechanics*. Based on *Schrodinger's wave equation* relating the energy of a system to its *wave function*, only certain values for which are allowed.

Wavemeter. An instrument used for measuring the *wavelength* of a *radio frequency electromagnetic radiation.*

Wave Motion. The propagation of a periodic disturbance carrying *energy*. The locus of these displacements at any instant is called the wave. The *wave* motion moves forward a distance equal to its *wavelength* in the time taken for the displacement at any point to undergo a complete *cycle* about its mean position.

Wave Number $\sigma = 1/\lambda$. The number of waves in unit length. It is the reciprocal of *wavelength.*

Waves-Audible. Waves with frequency ranging between 20 and 20000 vibrations per second, these waves can be heard by human ear.

Wave Theory of Light. Refers to the theory that *light* is propagated as a *wave motion.*

Wave Train. Series of waves of limited duration produced by same disturbance.

Wax. Type of lipid useful in protection of organism ; non-greasy, insoluble solid substance that melts at fairly low temperature paraffin wax; used for sealing, making candles and by shoemakers.

W-band. Band of microwave frequencies ranging from 56 to 100 gigahertz.

WBCs (White Blood Corpuscles). Leucotes ; colourless blood cells present in vertebrates.

Weak acid. It is not completely ionised in solution.

Weak Interactions. A fundamental interaction between *elementary particles* that is some 10^{12} time weaker than *strong interactions. Beta decay* is a form of weak interaction.

Weather. Atmospheric condition ; sum total of abiotic factors like sunlight temperature, cloudiness, wetness, wind and pressure.

Weber. The *SI unit of magnetic flux.* It is the flux that, linking a circuit of one turn, produces in it an *E.M.F.* of one *volt* as it reduces to zero at a uniform rate in one *second.* Symbol Wb (=Vs). 1 weber = 10^8 *maxwels.*

Weight. The *force* of attraction of the *Earth* on a given mass, weight is correctly measured in *newton.* The weight of a mass *m*, is equal to *mg*, where *g* is the *acceleration of free fall.*

Weight, British Unit Of. Avoirdupois weights.

$437\,^1/_2$ grains =1 ounce
7000 grains = 16 ounce = 1 *pound* =0.453592 *kilograms*
14 pounds = 1 stone
2 stone = 1 quarter
4 quarters = 1 hundred weight
2240 pounds = 20 cwt = 1 ton
2000 pounds = 1 short ton.

Weight, Metric Units Of.

1000 milligrams = 1 gram=15.432 grains
1000 grams = 1 kilogram = 2.20462 lb
1000 kilogram=1 tonne = 0.984207 ton.

Weightlessness. Refers to the state of a body which is infinitely far from any other body so that it experiences no *gravitational force.*

Welding. Process of joining of two *metal* surfaces by increasing their *temperature* very high so as to melt and fuse them together.

Weston Cell. Cadmium cell. A *primary cell* used as a standard of *E.M.F.* It produces 1.0186 *volts* at 20°C.

Wet and Dry Bulb Hygrometer. An instrument used for determining the *relative humidity* of the *atmosphere.* It consists of a pair of *thermometers* side by side, the bulb of one being surrounded by moistened muslin.

Wet Gas. Natural gas containing liquid hydrocarbons.

Wetting Agent. Substance that lowers surface tension of liquid.

Whale-oil. Animal fat obtained from fatty layer of blubber of whales.

Wheatstone Bridge. A divided electrical *circuit* used for the measurement of *resistance.* This principle is applied in the metre bridge. A wire, *AB,* of uniform resistance and generally I metre in length, corresponds to R_3 and R_4 in the Wheatstone bridge diagram; for R_1 a standard resistance is used, while R_2 is the resistance to be measured. By a sliding contact a point of no deflection in the galvanometer is found along *AB,* the resistance R_3 and R_1 being proportional to the lengths cut off.

Whistler. An *atmospheric* whistle of descending *pitch* which can be picked up under certain circumstances by a *radio* receiver. It is due to *electromagnetic radiations*, produced by *lightning* flashes, which follow the *lines of force* of the Earth's *magnetic field* and are reflected back to Earth by the *ionosphere.*

White Arsenic. Arsenious trioxide (As_2O_3); white crystalline, active poison containing 75% arsenic; used as insecticide in poison baits, being phytotoxic it is not applied on plants.

White Bronze. *Bronze* containing a high proportion of tin.

White Dwarf. A class of small highly dense *stars* of low *luminosity.* They have high surface *temperatures* and appear white.

White Glass. Sodium silicate (Na_2SiO_3); soluble crystalline salt; used in fire proofing textiles and in manufacture of cement and paper.

White Lead [Pb $(OH)_2 2PbCO_3$]. Basic lead carbonate white powder obtained by action of acetic acid vapour, moisture and carbon dioxide upon lead; used as white paint.

White Light. Light consisting of a continuous range of wave-lengths, mainly in visible range.

White Spirit. Mixture of alkanes having boiling range of 150 to 200°C; used as solvent and in paint and varnish industries.

White Vitriol. Zinc sulphate ($ZnSO_4$).

Wide Angle Lens. Camera lens with angle of view generally greater than 80° and short focal length.

Wiedemann-franz Law. The ratio of thermal conductivity to electrical conductivity is the same for all *metals* at a given *temperature*. This ratio is proportional to the *thermodynamic temperature*. Most pure metals obey the law at ordinary temperatures.

Wien Displacement. The product of the *thermodynamic temperature* and the wavelength at which maximum emission occurs from a *black body* is constant.

Wigner Effect. The effect produced when the *atoms* in a *crystal* are displaced as a result of *irradiation.*

Wigner Energy. Energy stored within a crystalline substance as a result of the *Wigner effect.* In a *nuclear reactor* in which *graphite* is used as the *moderator*, some of the energy lost by the *neutrons* is stored in the graphite; which is known as the Wigner energy.

Wigner Nuclides. Pairs of *isobars* of odd *mass number* wherein the *atomic number* and *neutron* number differ by one, *e.g.* $^3/_1$H and $^3/_2$He.

Wimshurst Machine. A laboratory apparatus used for generating static *electric charge.* It consists of two insulating discs to which radial metal are attached. The discs are rotated in opposite directions. The charge, produced by friction, is collected by metal combs.

Wind. A large-scale movement of air, caused by a *convection* effect in the *atmosphere.*

Window 1. A wavelength band to which a particular medium is transparent. The atmosphere, for example, has a radio window in the range 8 mm—20m. 2. A period during which an event may take place. For example, a spacecraft may be launched during a launch window to achieve a desired rendezvous.

Witherite ($BaCO_3$). White mineral barium carbonate.

Wolframite (WO_4). Mineral ore of a tungsten.

Wollaston Prism. Polarising prism; used to obtain plane polarised light.

Wood Alcohol. Methanol (CH_3OH).

Wood Metal Alloy of bismuth (50%), lead (25%), tin (12.5%) and cadmium (12.5%).

Woofer. Loudspeaker designed to reproduce lower audio-frequency sounds.

Word. The smallest number of *bits* of information that a particular *computer* can conveniently process as a single unit ; usually 12 to 64 bits.

Word Processor. A computerized typewriter consisting of a keyboard, a *microcomputer* a memory store, and a *VDU*.

Work (Phys.) The work done by a *force f* when it moves its point of application through a distance *s* and is equal to $fs \cos \theta$, where θ is the *angle* between the line of action of the force and the displacement. The *SI unit* of work is the *joule*; other units are *erg, foot-pound, foot-poundal.*

Work Function 1. The work function of a metal is the energy that must be supplied to free electrons possessing energy *E*, to enable them to escape from the metal. It is usually expressed in joules **2.** Helmholtz *free energy* (*A*).

Woulfe's Bottle. Glass bottle having two necks ; used for passing a gas through a liquid.

Wrought Iron. Purest commercial form of iron.

Wurtz-Fitting Reaction. Chemical reaction wherein aromatic halide reacts with alkylhalide in presence of ether and sodium to form aromatic hydrocarbons.

Wurtz Reaction. Method of synthesising hydrocarbons by treating an alkyl halide with sodium in ether.

X

Xanthate. Salt of xanthic acid.

Xanthene. While crystalline compound of carbon, hydrogen and oxygen: used as dye-stuff.

Xanthic Acid. Series of additional compounds of alcohol with carbon disulphide.

Xanthin. Insoluble yellow colouring matter found in various flowers.

Xanthine ($C_6H_9O_2N_4$). White amorphous powder formed during conversion of urea to uric acid in liver of uricotelic animals ; also present in tea leaves and meat extract.

Xanthone ($C_{13}H_8O_2$). Colourless crystals, obtained by action of heat on phenyl-salicylate.

Xanthophyll ($C_{40}H_{56}O_2$). One of the two yellow pigments found in chlorophyll of green plants.

Xanthophyta. Yellow-green algae.

X-band. Band of radio frequencies extending from 5200 to 10900 megahertz.

Xenon (Xe). Colourless monoatomic gas of zero group, occurs in minute amount in air (0.00001%); used in certain types of thermionic valves, light bulbs and fluorescent tubes.

Xerography. Non-chemical (dry) method of photography or photocopy; image of the material is electrically charged on a surface (paper) and developed by dusting with electrically charged fine dry ink powder.

Xerophthalmia. Drying of retina due to deficiency of vitamin A.

Xeroradiography. X-ray photography with the help of xerography.

Xerostoma. Excessive dryness of mouth caused by saliva deficiency, often causing dental decay.

Xerox. Device for copying printed matter etc., by xerography.

X-radiation. Pertaining to electromagnetic radiation of frequencies between that of ultraviolet radiation and gamma rays.

X-ray crystallography. The study of crystalline substances by observation of the *diffraction* patterns which occur when a *beam* of *X-rays* is passed through a *crystal*. Structure of certain *protein (e.g., haemoglobin)* and *nucleic acids* has been analysed using this techniques.

X-ray(s). Electromagnetic rays of very short wave-length, they can penetrate matter opaque to light rays; they pass through mammalian flesh while bones are opaque to such rays. Discovered by Wilhelm von Rontgen in 1895.

X-ray Spectrum. Each *element*, when bombarded by *electrons*, emits *X-rays* of several characteristic *frequencies*, depending on the *atomic number; spectrum* corresponding to various elements can be obtained from the X-rays emitted.

X-ray Tube. An evacuated tube for producing *X-rays*.

Xylan. Complex polysaccharide found closely associated with cellulose in plants.

Xylene[$C_6H_4(CH_3)_2$]. Xylol; dimethyl benzene; liquid found in coal-tar, boiling point 137°C; used in manufacture of dyes.

Xylenol [$(CH_3)_2C_6H_4OH$]. Refers to any of six monohydric phenols derived from xylenes.

Xylidine [$(CH_3)_2C_6H_3NH_2$]. Dimethyle aniline; aromatic amine, found in six isomeric forms, boiling point 216 to 230°C ; used in manufacture of dyes.

Xylose ($C_5H_{10}O_5$). Wood sugar ; colourless, crystalline pentose, found in xylane.

Xylyl. Univalent radical, $CH_3C_6H_4CH$.

Xylylene. Bivalent radical, $—H_2CC_6H_4CH_2$.

Xyster. Instrument used for scrapping bones.

Y

Yagi Aerial. A directional *aerial* used in *television* and *radio astronomy*. Named after Hidetsuga Yagi (1886-1976).

Yard. British unit of length. The yard as defined by the 1963 Weights and Measures Act is 0.9144 *metre.*

Year. A measure of time, the time taken by the *Earth* to complete its *orbit* found the *Sun.*

Yeasts. *Unicellular micro-organisms* producing *zymase*, they convert *sugars* into *alcohol* and *carbon dioxide*, used in *brewing* for the production of alcohol, and in baking.

Yield Point. It is the point at which a sudden increase in elongation occurs with only a small increase in tension.

Yogurt (Yoghurt, Yoghout). Thick fermented liquor obtained from milk; curd.

Young's Modulus. Elastic modulus applied to a stretched wire or rod under tension or compression ; ratio of longitudinal stress on cross-section of wire or rod of longitudinal strain.

Yow. Female Sheep.

Yperite. Musturd gas.

Ytterbium (Yb). Soft silvery element of lanthanoid series, boiling point 1194°C; used for improving quality of steel.

Yttrium (Y). Silvery-white metallic element, of transition series, boiling point 3337°C ; used in making alloys.

Yukawa Force. Strong, short-range force between nucleons; calculated on assumption that this force is due to exchange of a particle of finite mass (yukawa meson).

Z

Zeeman effect. Refers to splitting of spectral lines in radiation emitted by atoms or molecules in static magnetic field.

Zeeman energy. Interaction energy between atomic or molecular magnetic movement and applied magnetic field.

Zein. Simple protein with low molecular weight, found in maize.

Zener Current. The current in a *semiconductor*, made up of *electrons* that have escaped from the *valence band* into the *conduction band* due to a strong *electric field.*

Zener Diode. *Semiconductor diode* of a *p-n* junction working as a *rectifier* until the Zener breakdown voltage is reached, the device then becomes conducting due to the *Zener current.*

Zenith (astr.) The highest point; the point on the *celestial sphere* directly overhead.

Zeolites. A class of *hydrated aluminosilicates*, both natural and synthetic, used as *ion exchangers* and as absorbents, called molecular sieves.

Zero. Nought; the starting-point of any scale of measurement.

Zero Gravity. Weightlessness, in which no acceleration whether of gravity or other force can be detected.

Zero Nerve. Terminal nerve coming out from ventral side of cerebrum through neuropore and going into olfactory mucous membrane.

Zero Point Energy. The *energy* possessed by the *atoms* or *molecules* of a substance at the *absolute* zero of temperature.

Zerovalent. Having a *valency* of zero.

Zeta. Zero Energy Thermonuclear Apparatus. A *torus*-shaped apparatus for studying controlled *thermonuclear reactions.*

Ziegler Catalysts. *Catalysts* capable of initiating the *polymerization* of *ethene* and *propene* at normal *temperatures* and *pressures, e.g.* titanium trichloride and aluminium alkyl.

Zinc (Z). Hard bluish-white metal, melting point 420°C, boiling point 907°C; used in alloys and in coating iron sheets and wires to protect against rusting.

Zincate. A *salt* containing the *ion* ZnO_{22}^{-}.

Zinc Blende. Sphalerite. Natural *zinc sulphipe*, ZnS. *ore* of zinc.

Zinc Carbonate. *Calamine*. $ZnCO_2$. A white *insoluble* crystalline substance, used in medicine for treatment of skin diseases.

Zinc Chloride ($ZnCl_2$). Deliquescent white solid, soluble in water forming acidic solution; used as dehydrating agent, for preserving wood and making paste with zinc oxide, used as dental cement.

Zinc-copper Couple. Metallic zinc with a thin copper coating. It evolves hydrogen with hot *water*.

Zincite. Natural *zinc oxide*, ZnO. *ore* of zinc.

Zinc Fluorsilicate ($ZnSiF_6.6H_2O$). Water soluble salt; used as wood preservative and as fungicide.

Zincite. Natural zinc oxide (ZnO), an important ore of zinc.

Zinc Oxide. ZnO. A white *amorphous* powder, m.p. 1975°C., used as a *pigment*, in *glass* manufacture, in cosmetics, and in medicine.

Zinc Phosphide. Zn_3P_2. A grey *insoluble* crystalline substance, used as a rat poison.

Zinc Silicate. Various silicates of zinc exist. Natural zinc silicate, hemimorphite, is $2ZnO. SiO_2. H_2O$. Zinc *meta* silicate, $ZnSiO_3$, has m.p. 1470°C. Zinc *ortho* silicate, Zn_2SiO_4, has m.p.1509°C.

Zinc Sulphate ($ZnSO_4$). White crystalline solid, soluble in water prepared by roasting sulphide ore with air at moderate temperature; used in white paint, zinc-plating etc.

Zinc sulphide (ZnS). White amorphous material prepared from solution of zinc salts; found in nature as zinc blende; used as pigments in paints.

Zircon. Zirconium silicate. $ZrSiO_4$. A colourless *insoluble* substance, m.p. 2550°C., used as a gemstone when *transparent* and a *refractory* when coloured.

Zirconia. *Zirconium dioxide.*

Zirconium. Zr. Element. R.a.m. 91.22. At No. 40. A rare greyish *metal*, r.d. 6506, m.p. 1852°C., b.p. 4377°C. used in *alloys*, abrasives, and flame proofing *compounds*.

Zirconium Dioxide. Zirconium (IV) oxide, zirconia. ZrO_2. A white crystalline *insoluble* substance, m.p. 2715°C., used as a *pigment* and a *refractory.* The *hydrated* form, $ZrO_2 . xH_2O$. known as 'zirconium hydroxide' and 'zirconic acid', is a white *amorphous* powder.

Zirconium Silicate ($ZrSiO_4$). Colourless or slightly yellowish substance, insoluble in water; used as gemstone when transparent and refractory when coloured.

Zirconyl. The *univalent* group ZrO —

Zodiac. The zone of the *celestial sphere* containing the paths of the Sun, the Moon, and the Planets. It is bounded by two circles, which are equidistant from the *ecliptic* and about 18° apart. It is divided into the 12 signs of the zodiac, which are named after the 12 constellations.

Zodiacal Light. A faint luminous patch visible in the sky, on the western horizon after sunset or on the eastern horizon before sunrise. It is due to the *scattering* of sunlight by dust particles revolving around the *Sun.*

Zona Radiata. Zona radiata pellucida that exhibits radial striations.

Zone of Sphere. A portion of the surface of a *sphere* cut off by two parallel *planes.* Its area is $2\pi rd$, where r is the radius of the sphere and d the distance between the two planes.

Zone Refining. A purification method, used for *alloys, metals*, and *semiconductors,* based on the principle that the solubility of an impurity B in a main component A in the solid state may differ from the solubility of B in A in the liquid state.

Zones, Fresnel. *Half period zones.*

Zones of Audibility. An intense *sound, e.g,* due to an *explosion*, can usually be heard or detected at all points in a large area around the source of the sound, and also in distant zones of audibility separated from that area by regions in which the sound cannot be detected. Sound waves can reach these zones by reflection down from the *upper atmosphere.*

Zonula. Fibres holding crystalline lens of eye in position.

Zoochemistry. Chemistry relating to animal body.

Zoom Lens. System of lenses used in movie camera wherein focal length can be adjusted continuously to vary magnification without loss of focus keeping the image in same image plane.

Zorgite. Metallic copper-lead selenide.

Zwitterion. Dipolar ion that carry both positive and negative electric charge.

Zymase. Enzyme present in yeast which acts on sugar with fermentation of alcohol and carbon dioxide.

Zyme. Ferment; disease-germ.

Zymogen. Non-catalytic substance secreted by protoplasm of animal and plant cells, as a stage in development of an enzyme, proenzyme.

Zymology. Science of fermentation.

Zymometer. Instrument used for measuring degree of fermentation.

Zymophore. Active portion of enzyme.

Zymotic. Caused by fermentation or infection.

Zymurgy. Technological chemistry dealing with brewing, distilling and wine making *i.e.*, processes involving fermentation.